The Wisdom of Water

Alanna Moore

Python Press

Alanna Moore is the author of
'Backyard Poultry – Naturally' 1998
'Stone Age Farming' 2001
'Divining Earth Spirit' 1994 & 2004
'The Magic of Menhirs & Circles of Stone' 2005
'The Wisdom of Water' 2007
'Water Spirits of the World' 2008
'Sensitive Permaculture' 2009

Enquiries to Python Press
PO Box 929 Castlemaine
Vic 3450 Australia
www.pythonpress.com
pythonpress@gmail.com

'The Wisdom of Water'
is published by Python Press
ISBN – 978-0-9757782-1-0

Text and photographs by Alanna Moore.
Copyright © Alanna Moore, July 2007

Front cover: Pine Lake, Tasmania

Printed by Lightning Source

Dedicated to the Spirit of Water,
with thanks for help and inspiration to:
Rob Gourlay, Tim Strachan, John Archer, Morad Eghbal,
Billy Arnold, Alex Martin, Parvati, Lee and Frits Ringma,
Natural Resonance Study Group, Laurie Adamson,
The Edwards family, Pauline Roberts, Richard O'Neill,
Janet Millington, Maureen Brannan, Francis Wade,
Jo Dean, Ilyhana Kennedy.

All rights reserved. No reproduction, copy or transmission
of this publication may be made without written permission,
in accordance with the provisions of the Copyright Act.

The Wisdom of Water
Contents:
Introduction   7

Part One: Waters of the Earth

1.1 Water and the Rise and Fall of Civilisations   11
Water and civilisations – Degraded waters – Vanishing waters and the Holy Grail – Tree loss and rainfall – Plantations dry out the land too – The Cauldron of Rebirth.

1.2 Water in Australian landscapes   20
De-watering of Australia's landscapes – Victoria's vanishing rivers – Aboriginal water harvesting in Australia – Chains of ponds – Ancient Aboriginal gardens – Eating carp – Sustainable Aboriginal aquaculture – The wealth of swamps – Water for the Wimmera-Mallee.

1.3 Ground water   33
Groundwater origins - Groundwater knowledge in Australia - How do they find 'new water'? - Australia's Great Artesian Basin - Desert oases - Unleashing the unthinkable? - Hot springs in Australasia - Australasian mineral waters - Wetlands and radioactivity.

1.4 Water Divining   67
Dowser finds underground river – No surface water? Try water divining! – How do dowsers find water? – What water diviners look for – Dowsing water depth – Dowsing water quality – Some pitfalls of water divining – Dam leaks? – Well gone dry? - *Village Water*: a dowser's charity – Dowser society contacts.

Part Two: Waters of power and mystery

2.1 Water's special qualities — 81
What is water? – Water as a crystal – Water loves curves – Water and energy – Sun and water makes natural energy – Temperature differences – Energies of thunderstorms – Lightning and life – Storms and ionisation – Water as a geopathic irritant – Water's memory – Water and emotions – Surface tension.

2.2 Water and phenomena — 93
Mystery lights – Water and UFOs – Water energies and dowsing.

2.3 Tapping water's wisdom — 97
Water and wisdom – More on water and emotions – Watery transformation – Attracting water – The gift of water – Tears of the Earth – Eye wells – Well heads – Oracle waters – Springs and wells predict earthquakes and rain – Well dreaming at Glastonbury – Continuing oracle traditions.

2.4 Water Worship — 115
Spirit of water – Amphibious gods – Storm gods – Water goddesses – Queen of the South Seas – Water shrines – Sacred rivers, lakes and springs – Sun and water – River goddesses – Magic and healing waters – Healing properties – Balinese water goddess – Water and spirit in the Bible – Angel of water – Water and purity – Lustration – Churches and wells – Well pilgrimages – The water goddess today.

Part Three: Waters of the Sky

3.1 Droughts in Australia   143
Diary of a drought – Droughts, politics and deja vu – What of El Niño? And La Niña? – Climate change predictions – What about climate cycles? – Soil and rainfall – Droughts are followed by floods – Drought and fire – Smoke and air pollution reduce rainfall – Drought and frost – Dust storms – Sinking and shrinking – Drought and wildlife – Toxic soup – Water theft – How much water is there? – Where does our water go?

3.2 Weather prediction by nature   159
It's all in the flowering – Animal behaviour – Aboriginal weather wisdom – Cockatoos know! – Indian folk wisdom – Astronomical prediction – Prediction by dowsing.

3.3 Rain-making traditions   167
Universal rituals – Chinese water dragons – Rain-making in Africa – Rain-making in India – Rain-making in Europe – Rain-making in Australia – The Nyigina way – Rain-making art – Rain songs – Rain-making in New Zealand.

3.4 Sky water harvesting   177
Outer space origins – Rivers in the sky – Water vapour and condensation – Air wells – Dew-ponds – Fog fences – Let the people drink thin air! – Water from wind – The Spray Turbine – Modern rain-making – Energy balancing and rainfall – 'Cloud-busting' results – Rain-making in biodynamic farming – Prayer and rain invocation – When dowsers prayed for rain – Geomantic rain-making – Rain songs re-energise the land.

Part Four: Restoring the Waters

4.1 Re-watering Australia's landscapes 197
Grassroots solutions to soil salinity - Science at the grass roots - Useless government advice - First interceptor banks - Encouraging results - Benefits of interceptor banks - A dowser's innovations - One man's interceptor banks - What has science to offer? - Exposing salinity myths - Restoring the land's water balance - Re-creating chains of ponds - Saving soil water on the farm - Keyline systems - Water-friendly ripping and ploughing - Permaculture swales - Ponding banks in Central Australia - Sustainable pastures - Stormwater harvesting in the city.

4.2 Household water conservation 216
Low-water gardening - Grey water in the garden - Urine in the garden - Low-water sustainable diet.

4.3 Improving water quality 223
Contaminated river waters - The glow in your glass - Endocrine disrupting chemicals - Where river deltas meet the sea - Who wants to use recycled water? - Chemical cocktail, anyone? - Testing the waters - Groundwater contaminants - Problems of rainwater - Water storage and conveyance - Collecting rainwater - Acid or alkaline water? - Filtering and sterilising water - Oxygenated water - Solar water purifier - Desalination and wave energy - New wave of water conditioners - Flowforms - Energising water with crystals - Magnetising water - The power of prayer - A dowser cleanses water.

4.4 Water wisdom today 292
Developing a spiritual relationship with water - People power at work - Walking the Wimmera - Healing the waters - Agnihotra ritual cleanses a sacred spring - World day of love and thanks to water - Tibetan Buddhist water ritual - Many ways to care for water.

Introduction

Water is everywhere. It covers about 70 percent of the planet. Yet just about everywhere water is becoming more scarce. Or is it? Perhaps we are not looking for new water supplies in the right places.

Water is everywhere, yet it is miraculous and mysterious. Revered since earliest times, in the current era we have been abusing it and taking it for granted for too long. Certainly, when it arrives out of a tap a lot of its original magic is gone. But it can be rescued!

Water also constitutes around 70 percent of our bodies (even more so in babies and our brains). So we are microcosmic oceans ourselves, pulsating in resonance with the rhythms of life; our bodies responding well to hydrotherapy and water based vibrational medicine.

I have been writing this book in a time of severe water crisis in Australia, when the stress on surface waters has never been so extreme. We have had serious droughts before, but never have we continued extracting water at such a rate. The damming of some of Australia's last wild rivers (in Queensland) is under way – a desecration of wilderness and some of the last homes of the endangered lungfish. And nuclear powered desalination plants seem to be lurking just around the corner, if we choose to believe in the spin that surrounds them.

There must be better ways to harvest the waters of the Earth without destroying our environment in the process. Actually, there are! This book focuses on water in the Australian context (and beyond), its various manifestations, unique characteristics and abilities, and especially its folklore and spiritual dimensions. For the author is a professional geomancer, one who works with the subtle dimensions of place.

The geomantic perspective is that water is a conveyor of the intelligence and vitality of the planet, a supreme repository of creative love-wisdom. If we accept this view, then our lives are all the richer for it. For we are water too!

I want to take you on a journey to the heart of water, to dive into the deeper levels of water's subtle dimensions, to where new holistic understandings flow. You will not remain unchanged!

Part One:

Waters of the Earth

1.1 Water and the Rise and Fall of Civilisations

Water and civilisations

Throughout history, the rise and fall of great civilisations has hinged on the bounty and decline of plentiful fresh water supplies. Sedentary lifestyles were made possible by innovations in agriculture, with harvests as reliable as the rains and springs that nourished them. Many ancient cities owe their continued existence to the freshwater springs that sustain them; while the mechanisms of the Industrial Revolution were often powered by water mills.

In times past the importance of protecting and avoiding pollution of the life-giving waters was obvious to all. Legends of fearsome water monsters, such as dragons and bunyips, helped to keep people a healthy distance away from the all important water sources, as well as reducing the potential for drownings. Nobody lived in flood zones and swamps.

Australian Aboriginal people kept a healthy distance from waterways and wetlands and tended to live more on the ridges, where it was drier and not flood prone. They went to the rivers for food harvesting and eating; and they buried their dead, often, on river banks, bends and junctions.

But understanding and respect for landscape hydrology was swept aside when colonial cultures invaded and wrenched control of places such as Australia and Africa. The land so cherished by the indigenous people was then viewed completely differently, as a resource to be ravaged, its people brutalised and enslaved.

There were practical reasons to keep a safe distance from water. In tropical Africa, malaria has not always been such a problem as it is today. African people historically made their settlements on ridges rather than in valleys and thus avoided much contact with mosquitoes. When European colonisers arrived, cities were established on rivers that had become important transport corridors for the plundered goods and resources that were shipped back to Europe. Malaria became rampant.

In the 19th century, as the worlds big cities bulged with high populations, high death rates were curbed when understanding about water borne diseases and good sanitation, including the provision of sewerage, allowed exponential growth of the human population. The rivers became dumping grounds for our filth.

But to this day there remains a huge portion of Earth's population – several billion – who suffer contaminated water supplies and deadly disease outbreaks, especially in drought ravaged parts of Africa. The World Health Organisation has estimated that over 5 million people each year die globally from drinking contaminated water.

European settlement has been a disaster for Australia. Colonists exercised a serious lack of understanding when they drained wetlands to create more pasture or crop lands. They destroyed the land's natural water balance. Soil fertility rates on farmland have been on a downward spiral. Propped up nowadays by artificial fertilisers, which introduce more problems, Australia's agriculture today is rarely sustainable.

Wetlands continue to be reclaimed along the world's over-populated coast lines, as people seek the 'sea-change' lifestyle. Marinas and canal-side housing estates not only destroy important hydrological features, but also expose residents to geopathic stress, caused by spending time in 'irritation zones', where water is flowing under the ground beneath them.

Degraded waters

The domination of monotheistic religion spelled the decline of animist beliefs which had helped preserve sacred watery environments. Rivers, once revered as gods and goddesses, were doomed to be merely conveyors of transport and sewage.

Excessive deforestation caused siltation of the waterways and, with the loss of natural riparian vegetation from river banks – as Austrian 'water wizard' Victor Schauberger observed in the early years of the twentieth century – water became too overheated to be healthy and lost it's vital energy.

Nowadays many rivers are so toxic with industrial waste that little life remains in them. Raw sewage is discharged into rivers in many parts of the world. Constricted within rigid concrete canals and criss-crossed by metal bridges, many are no longer recognisable as real rivers any more.

Two major rivers are a case in point – the River Ganges in India and the Yellow River in China. Professor Jeffrey Sachs, director of the United Nation's Millennium Project, told an environment conference in New Delhi in January 2007 that these rivers are no longer flowing, they are pretty much stagnant from siltation and water extraction upstream, and the Ganges is seething with faecal coliforms from inadequately treated sewage.

Humans are very good at fouling their nests and poisoning their sources of sustenance. Sanitation laws did reduce disease outbreaks and made the urban world more pleasant, but they don't address fundamental problems of human water use. In modern Australia and elsewhere strict planning laws keep household waste water disposal at optimal safe distances from waterways to prevent pollution. But they do not put into the equation the inappropriateness of mingling waste with water, here on the driest of continents. The morality of our society's relationship with water begs questions. How can we justify using flush toilets in a land of regular droughts? Dry composting toilets are so much more eco-friendly!

Vanishing waters and the Holy Grail

In the medieval romance of the Holy Grail, a priestess of a sacred spring was violated – raped by an evil English king, who stole her golden cup, the symbol of her benelovence. The King's men followed suit and after that, the priestesses of the holy wells no longer welcomed wayfarers with food and drink. As a result, the waters retreated back into the Earth, withdrawing their fertility. The land went to waste and there was illness and famine.

Fortunately a good knight found out about the Grail's ability to restore the waters and so began King Arthur's quest to find it. When the Holy Grail was recovered, the waters returned and the land came back to life.

That an offence to Mother Nature could inflict barrenness on the land was a recurring theme in many medieval tales. The idea runs deep. Some people attribute the desolation of the north African and Arabian

> environments to Islam's suppression of earlier Goddess traditions. For it was the triple aspected moon goddess who was originally worshipped at the black stone, the Kaaba, in the Haram sanctuary of Mecca. A sacred spring is located nearby – the Zamzam. In legend, Abraham's wife Hagar was said to have nursed Ishmael, founder of the Arabian people, back to health with the healing waters of the Zamzam spring – the source of one of the 'rivers leading to Paradise' in Islamic tradition. The region was dotted with sacred stones and hills and was a pilgrimage centre long before Islam.

Tree loss and rainfall

Over ten millenia have passed since the great agricultural experiment was begun in Eurasia. Many ancient centres that were birthplaces of civilisation and reliant on irrigated agriculture, now lie parched and sunbaked in desolation. Climate-modulating forests have been reduced to pasture. Mountains of waste and poisons pollute the land and waters. Mother Gaia must dread mankind's footprint.

Intensive agriculture plus changes to the world's climate have often combined to spell disaster. Evidence suggests that North Africa and the Middle East have been experiencing a 5000-year trend of reducing rainfall and desertification. The Middle East has now lost much of its fertile top soil, the cedars of Lebanon are mostly gone. Between the mighty Tigris and Euphrates rivers, the once 'fertile crescent' is now arid and brittle, and vast stretches of sub-Saharan Africa, northern China, the Burren in Ireland and endless other vast tracts of farmland, are now inhospitable.

Twenty three hundred years ago Plato bemoaned just such a calamity in southern Greece with astute insight, saying that:

'At the period, however, with which we are dealing, when Attica was still intact, what are now her mountains were lofty, soil-clad hills; her so-called shingle plains of the present day were full of rich soil; and her mountains were heavily afforested – a fact of which there are still visible traces. There are mountains in Attica which can now keep nothing but bees, but which were clothed, not so very long ago, with fine trees producing timber suitable for roofing the largest buildings; the roofs hewn from this timber are still in existence. There were also many lofty cultivated trees, while the country produced boundless pasture for cattle.

'The annual supply of rainfall was not lost as it is at present, through being allowed to flow over the denuded surface into the sea, but was received by the country, in all its abundance, into her bosom where she stored it in her impervious potters earth and so was able to discharge the drainage of the heights into the hollows in the form of springs and rivers with an abundant volume and a wide territorial distribution. The shrines that survive to the present day on the sites of extinct water supplies are evidence for the correctness of my present hypothesis.'
(From Plato's Critias.)

Time after time farm lands have been milked of their richness, laid to waste and abandoned around the world. Mankind should know better by now, but still the fertile soils are subject to abuse and the amount of arable land is shrinking. The current chronic lack of rain in regions such as southern Australia cannot simply be the result of climate change plus natural cycles of drought. Like Plato's Attica, our land's moist abundance has been despoiled by massive amounts of tree clearing, followed by over-intensive and inappropriate farm practices.

Australia has a terrible land clearing track record. In just one year (2001-2002), farmers in Queensland, fearful about impending new anti-clearance laws, frenetically wiped out 1.1 million hectares of native trees, it was revealed satellite photography. Unsurprisingly, in 2002 a national audit of Australian ecosystems by leading scientists stated repeatedly that 'vegetation clearing is the most significant threat' to the survival of Australia, closely followed by overgrazing. Nobody seemed to care much. The report languished for six months on the desk of the federal agriculture minister and had to be leaked to the media before it was finally released.

In 1994 another report was suppressed. This one revealed that 25,000 hectares of land across 41 shires in Western Australia were being illegally cleared annually, threatening the very existence of 4000 plant species. 'Farmers are clearing to replace land that has been lost to salinity' it was eventually reported in the West Australian. An embarrassment to the West Australian Department of Agriculture, Conservation and Land Management who had commissioned it, the report vanished and its author resigned in protest.

Where trees are abundant, rainfall is more likely to occur. And since

Plato's era, the connection has often been made between a reduction in tree cover and lessening rainfall. It has been a frequent observation and one need only ask 'old timers' in rural areas for such local ground truth perspectives.

Trees transpire moisture and help to encourage water table recharge. Anecdotal evidence points to a strong relationship between an appropriate amount of tree cover, good soil health and rainfall. But just how the trees grow is an important factor. Old growth forests with mostly mature trees, that have deep tap roots and minimal growth, make great water catchment areas. But plantations of trees, all uniformly young and growing fast, are very thirsty and can actually dry out landscapes.

Plantations dry out the land too

Geo-hydrologist David Leaman has been warning the Tasmanian public about this problem for several years. An industrial attitude to forestry has seen a rapid increase in monoculture eucalypt and pine plantations across the state, at the cost of magnificent natural forests. Leaman, in his book about water, tells us that:

'Land clearing and plantation development have significant results ... [which] began decades ago with pine plantings in northern Tasmania but have accelerated since 1991 and especially since 1997 with the use of Eucalyptus nitens. These trees are 'super suckers'.

'Plantations are now so widespread, increasing in area and especially concentrated in headwater segments of catchments – and in much of the northern regions – such that water supplies are being locally and regionally disturbed and flows distorted.'

The stress on catchments means plants and wildlife suffer, as well as people downstream. Leaman suggests conversion of many of the plantations to either just grassland, or back to forests of mixed native trees, such as slower growing hardwoods, as a way to restore the balance.

It's not just a problem for Tasmania either. Rob Belcher, chairman of Sustainable Agricultural Communities Australia, writing in an opinion piece (*The Weekly Times*, 9th May 2007) says that:

'Wherever I travel in Australia, the common complaint from landholders adjoining plantations is the reduction of stream flow and aquifer

> *resources, or the total loss of water ... The CSIRO and other experts calculate that one hectare of plantation will consume 1.5 megalitres of water per annum, over and above what would be consumed by unirrigated farm pasture.'*
>
> Belcher points the finger of blame at managed investment schemes (MIS) that have been fuelling the exponential and unsustainable growth of new plantations, giving investors 100 percent tax deductions to do so. An early 2006 tax office ruling has abolished the tax perk for agricultural projects, but tax on investments in plantation forestry are unchanged, threatening more of our precious water supplies. He says that:
> *[In 2006] ...' the MIS sector bought 80-85 percent of all tradeable water in the Murray Darling Basin ... and huge dams are being filled ... while the river struggles to flow .... This is an illogical, unsustainable, unfair, government sponsored water grab.'*

Whatever the cause and factors influencing the current drought, the future of Australia's great food bowl of the world is evaporating! After up to ten years of reduced rainfall in southern Australia, the 2006 harvest has been decimated and a new method of satellite mapping has found a net loss across Australia of 46 cubic kilometres worth of fresh water over the past three years. That's a lot of fresh water to vanish: enough to fill Sydney Harbour more than 90 times, the media reported at the end of 2006. But the climate change debate is so loud these days that it drowns most other observations and recommendations for environmental protection. Without an end to bulk tree clearance in places like Australia and the Amazon, and without a massive swing to sustainable agriculture, I imagine that decreasing rainfall and soil fertility is likely to continue.

'As long as bare land sheds water like a tin roof there will be ruinous erosion and serious flooding ... Dams are not a permanent answer ... The importance of getting rain water into the ground cannot be overestimated. This conserves both the water and the soil.'

So wrote an American author back in 1953. This statement still holds true. But it conflicts with entrenched scientific dogma in Australia, whereby we are supposed to be lowering the water table in order to reduce soil salinity – a totally erroneous theory.

Fortunately there is a wave of change beginning in this land. Just the other day I heard the great news that Tasmanian Forestry has finally decided to stop clear-felling majestic old-growth forests to establish tree plantations, as of 1st June 2007. They had brought forward their decision by some three years because of public pressure!

And it's not all doom and gloom on Australia's farms. Much degraded land is being restored to fertility with methods developed over the past 50 years by astute land managers who are keen to restore the water balance. Farm design systems such as Yeoman's keyline planning, permaculture design, Whittington's interceptor banks, plus the restoration of wetland areas; as well as Natural-Sequence, holistic, organic, biodynamic, sustainable low-input and biological farming regimes; combined with tree planting and the remineralisation of farm soils, are all helping to rescue land on the brink of collapse and recover its fertility.

The Cauldron of Rebirth

The legendary Cauldron of Rebirth, raised from a lake in Ireland, was a vessel capable of making the wounded and sick whole again. It symbolised the venerated womb of Mother Earth, wherein Her sacred waters percolated, and the gaining of salvation and rebirth from the Great Mother. It's successor became the Holy Grail. The Matthews write:
  [But] '...whether a cup or a cauldron, the essential healing element is the liquid which is contained by the vessel.'

We can dream of good soaking persistent rain, but to be realistic we also need to make do with what we've got and learn to use it better. Positive action is needed. We can cultivate that great Cauldron of Rebirth. Thanks to the wake-up call that the Australian drought provides us with – we can take action! Conserve what water there is and clean up our pollution. There is a lot that can be done by individuals, communities and governments. But a better understanding of water's many ways is needed.

References:

King, Thomson, *Water: Miracle of Nature*, Collier-Macmillan, USA, 1953.

Kininmonth, William, 'When politics engulfs science,' *The Age*, 8th July 2005.

Ramesh, Randeep, World is running out of water, says UN adviser in New Delhi, 22nd January, 2007, *The Guardian*, www.environment.guardian.co.uk/water/story/0,,1996211,00.html

www.sacredland.org/world_sites_pages/Mecca.html

Cunningham, Irene, *The Land of Flowers – An Australian Environment on the Brink*, Otford Press, NSW, 2005.

Flannery, Tim, *We are the Weather Makers*, Text publishing, Australia, 2005.

Matthews, Caitlin & John, *Celtic Wisdom*, Rider, 1994.

Alexandersson, Olof, *Living Water*, Turnstone Press, 1976.

Macey, Richard, 'The word from on high: we're drying up fast', *Sydney Morning Herald*, 30th December 2006.

Leaman, David, *Water – facts, issues, problems and solutions*, self published, 2004, Australia (available from Leaman Geophysics, GPO Box 320, Hobart, Tas 7000.)

1.2 Water in Australian landscapes

De-watering Australia's landscapes

In Western Australia early colonists reported on the abundant supplies of pure, sweet water in the south west of the young colony. In 1827 Captain James Stirling was the first white man to sail the Swan River and he went inland as far as Ellen Brook. This he described 'as a beautiful running brook watering several hundred acres of natural meadow, covered in rich green grass still, at the end of summer'.

But it's a very different story now at that time of year. 'Today in March the grass is bleached dry and there is no water in Ellen Brook' writes author Irene Cunningham. In 1939 Gingin Creek was said to be 'the life of the district', with tributaries fed from inexhaustible springs. Locals enjoyed long and healthy lives there.

Today the waters are undrinkable. Swimming in the Moore River is also a health hazard, thanks to siltation, nutrient overload from agriculture and the rot of eutrophication in hot weather and low flow situations. The story is repeated all across Australia, where only half of the original wetlands of 200 years ago remain.

In Victoria, like south western Western Australia, once known for its abundant seasonal wetlands, the picture is similar. Victoria is blessed with vast areas of rich basalt soils, so many wetlands here have been deliberately drained to create more fertile pasture or cropping land. Many of the wetlands that remain, such as Kow Swamp, Avoca Marshes and the Kerang Wetlands, are important wildlife havens which are listed as being of national and often international significance (often as Ramsar 1971 Treaty wetlands).

Victoria's vanishing rivers

When Surveyor-General Major Sir Thomas Mitchell explored Australia's south east in the 1830s, in his search for potential grazing and farming lands, he reported finding 'Australia Felix'! Enthusing about lush country around the Merino Tablelands, near present-day Casterton, he described it as where 'the hills swelled, the water foamed and glittered, the balmy air was sweetly perfumed, the grass was green as an emerald and covered with a thick matted turf'. It resembled a 'nobleman's park on a gigantic scale', he noted.

The graziers that came hot on his footsteps reported healthy rivers, some with deep permanent pools of up to some 9 metres (30 feet) deep around the Grampian Mountains (Gariwerd) region. Bridges were required to ford these rivers and huge Murray Cod fish could be caught in them, well up into the rivers' headwaters, such as where the Cairn Curran reservoir now dams the upper Loddon River.

By 1850 there were six million sheep in western and central Victoria. These soon stripped the vegetation, impacting heavily on the native soils and bulldozing the banks of waterholes. With their hard hooves, they contributed greatly to soil erosion and compaction.

Inappropriate land use policies of the 1830s, 1840s and later were a disaster for Australia's landscapes and especially her waterways. The replacement of native perennial grasses, such as kangaroo grass, by exotic annuals also contributed to erosion. From the 1860s, government policies for 'Selection and Closer Settlement' saw widespread tree ringbarking, land clearing, the drainage of swampland and frequent burning.

Rivers, generally not very common nor big in Victoria, were systematically cleared out and logs removed from them to speed up water flow. The settlers didn't have a clue about the consequences of this (nor of their agricultural methods). They failed to realise that the logs and snags provide important habitat for wildlife and that water turbulence around them creates deep holes, where fish can hide and thrive.

As a result, Victoria's once verdant rivers are now mostly shallow flat, often treeless and full of silt eroded from deforested areas. They flood easily, though, in Mitchell's time, they were never known to flood. Many, such as the intermittently flowing Avon-Richardson Rivers, are classed as having only poor or marginal condition these days. The situation is similar in the rest of the agricultural areas of southern Australia

Successful Casterton squatter John Robertson, wrote to Lieutenant

Governor La Trobe in 1853 about the demise of the already degraded Glenelg landscape, lamenting that:

'the long deep-rooted grass that held our strong clay hill together have died out; the ground is now exposed to the sun, and it has cracked in all directions; also the sides of precipitous creeks – long slips are taking trees and all with them. A rather strange thing is going on now. One day all the creeks and little watercourses were covered with a large tussocky grass, with other grasses and plants, to the middle of every watercourse but the Glenelg and Wannon, and in many places of these rivers, now that the only soil is getting trodden hard with stock, springs of salt water are bursting out in every hollow or watercourse, and as it trickles down the watercourses in summer, the strong tussocky grasses die before it, with all others. The clay is left perfectly bare in summer.'

Meanwhile the Murray-Darling basin has recently been classified as amongst the world's ten most endangered river systems. One of the world's longest rivers with one of the biggest catchments, the Murray's problems include flow regulation by dams and weirs, excessive water extraction, salinisation from run-off and climate change, says the World Wildlife Fund International (*Sydney Morning Herald*, 20th March, 2007).

Invasive fish and plant species are also worrying. Carp, for instance, constitute 60 to 90 percent of total river fish, with up to one carp per square metre. Carp are very destructive, digging up pond floors and riverbanks and bottoms in their search for food, while displacing native fish, and eating frog spawn etc.

Uncontrolled groundwater extraction in the Murray-Darling Basin is also impacting on river flows, with 'only 20–40 percent of major users of bore water metered', the Australian Bureau of Agricultural and Resource Economics announced (*The Weekly Times,* March 14th 2007). ABARE also conceded also that 'groundwater resources were hard to measure accurately'.

Aboriginal water harvesting in Australia

Living in a culture where cultivating food or keeping livestock was not practised in the European way, Aboriginal people's water requirements were not great. But even in the tropical north it can get very dry in the dry season. Waterholes were treasured all over. Tribal law maintained the all-important water sources in a pristine state and did not pollute them, and a fear of guardian spirits prevented much interference.

Soaks, small pools and springs were often covered with a rock, log or sticks to stop evaporation and fouling, or animals from drinking or drowning there. If they did get dirty, a hole would be dug beside the soak and the water that came up could be filtered through tufts of long grass, or the flower cones of Banksia trees would be placed between the lips and water sucked through them. This would strain out critters and muck, while adding aromas from the often abundant flower nectar, greatly improving any semi-stagnant flavour.

If rivers were dry, holes would have to be dug in the river bed, if the bottom was sandy, sometimes down to six metres (20 feet) down! Wells were always covered up again after use.

Down south in Victoria, where summers are dry, Aboriginal rock wells were used to collect and store water that ran off rocky outcrops. These were often natural water collection points on rock outcrops, and they were sometimes artificially enlargened with stone tools and fire. Like all waterholes, they could be associated with rainbow snake Dreamings and rain making ceremonies.

There are about 42 rock wells registered with Aboriginal Affairs Victoria and new ones are being re-discovered all the time. Some of the best preserved rock holes can be seen in the box ironbark forests in the Maryborough and Castlemaine regions of central Victoria, home of the Dja Dja Wrung clans.

The Whroo area, south of the Goulburn River, was home to the Nguraiillum-wurrung and Taungurong clans. A gold rush saw the township of Whroo established in 1850. It is now abandoned. All that remains of Whroo is a rock well, which local Aboriginal people continue to have an association with. This well, with its 110 litre capacity, ensured that the clan groups had a reliable source of water as they travelled through the area. When full, the water flows over a lip in the front wall. Whroo, Wahroo or Wooro means lips.

In south western Australia – *gnamma* is the Western Desert word for rock holes and this name has gone into common parlance elsewhere. The gnamma are usually found in granite outcrops and were important water sources for the people.

In the absence of surface water, dew would be shaken from plants and collected in a coolamon – a wooden dish, sometimes made from a hollow log end with its end stoppered up.

When drought times were really bad Aboriginal people in Victoria were sometimes forced to dig up frogs from their hibernation in the mud of dried-out creeks and squeeze the water out of them into their mouths. The roots of certain trees were also known to yield a drought-proof water supply.

You can imagine the horror when white people came and drank the gnammas dry and let their livestock trample the precious sacred springs and swamps. White people's appropriation of often meagre water resources was one of the most devastating impacts of colonisation.

Sometimes Aboriginal people were kidnapped and forced to disclose the whereabouts of local water supplies. At the beginning they had been happy to share their knowledge of water, until they saw the consequences of unfettered environmental abuse that was the white man's way.

Chains of ponds

North western Victoria was part of a huge inland sea some 25 million years ago. Drainage of watersheds went to the north as the sea receded. The Wimmera River, the only major watercourse in the region, begins in the hills between Maryborough and Avoca and travels through an ancient sandy plain to disappear into the ground at Lake Hindmarsh. First described by Major Mitchell in 1836, the river can be a series of pools or a raging torrent that connects the ancient river pathway in an overflow system known as Outlet Creek. This flows up through the centre of the Wyperfeld National Park, filling a chain of normally dry billabongs and former lakes and culminating at Lake Albacutya. Rarely do it's waters overflow via Albacutya to travel on through the Wirrengren Plain. It did so in 1918, the first time since 1852.

Mitchell, who insisted on finding out and using local Aboriginal names or words, named the river the Wimmera and went on to climb Mount Arapiles. From its summit he looked across to the north, to what is now the Little Desert National Park, and saw a network of shallow lakes in the slight depressions over impervious clays – 'an indication of better drainage and river systems in former times' writes Alan Fairly.

North of the Little Desert white explorers in 1844 described a large Aboriginal camp at an extensive lagoon that they learned was called Nhill. The name Nhill, they discovered, means 'abode of the spirits'. 'For they imagined that the mists rising from the water early in the morning were spirits of their ancestors,' writes Fairly.

During the gold rush, which started in the 1850s, a well-trodden track was made from South Australia to the goldfields. It wandered through the Little Desert between ephemeral springs, swamps and salt lakes, before crossing the Wimmera River at Polkemmet. No doubt the trail followed ancient Aboriginal travel routes.

An Aboriginal legend of the Wotjobaluk peoples of the Wimmera – Mallee regions tells of the formation of the Wimmera River and describes its chains of ponds beautifully. The lake system is said to be the giant Dreaming kangaroo Purra's footprints. This great ancestral kangaroo spirit hopped from Stawell to Gooro (Lake Hindmarsh) and Ngelbakutya (Lake Albacutya), marking the route of Barbarton, the Wimmera River. As Purra headed north his footprints became fainter and fainter, disappearing into the sandhills.

Three times last century Lake Hindmarsh dried up completely for several years. Floods, such as the exceptionally high floods of 1916 – 1917, bring the dried-out river systems to life, attracting water birds from far and wide to eat the aquatic life brought in by floodwaters.

The river overflow system was an important area for inter-tribal gatherings, where goods and produce were exchanged. It was the location of one of the largest campsites, at Wirrengren, in the middle of the plain, where gatherings were held at times of plentiful food supplies. Later at Wirrengren, the church-run Ebenezer Mission was deliberately built over a traditional ceremony site, which was also the site of a massacre.

A little to the north, in the Hattah Kulkyne National Park, the larger Hattah Lakes need a good Murray River flood every three or four years to not dry out completely. In such a flood event, the Murray's waters are diverted westwards via Chalka Creek. The larger lakes usually contain some water, but do dry up occasionally. A record flood level is marked up a tree beside Lake Hattah at 6.6metres (some 20 feet). The lake system filled up several times in the 1960s and fewer times since.

Aboriginal Latji Latji people made great use of these overflow wetlands, and there is evidence of their occupation in the form of middens of charcoal-blackened soils strewn with shells and bones alongside the channels. They erected weirs across the channels for fish trapping and collected mussels and yabbies (freshwater crayfish). The Latji Latji propelled themselves around their wet landscapes in canoes made from sheets of eucalyptus tree bark and pushed around with a pole. It must have been good living!

Ancient Aboriginal gardens

Around the Murray River and its tributaries there has been found evidence of Aboriginal gardens that are thought to be as much as 3000 years old. Artificially created mounds have been discovered that are up to one metre (three feet) high and 20 metres (60 feet) long. These garden mounds would have improved soil drainage and protected plants from frost in low-lying areas. They are usually found in clusters close to rivers, and over 3000 of them still exist in valleys, according to research by Dr Wendy Beck, of the University of New England's School of Archaeology and Palaeoanthropology, it was announced in 1998.

Over in western New South Wales at Brewarrina, the Nhunggabarra people maintained elaborate fish traps on the Darling River. An ingenious series of river stone weirs and pens were used to trap fish at various water levels. Fish migrated upriver to breed in flood times. As waters receded the big, old fish were smart enough to sense the danger and escape over the walls of the traps. Younger fish enjoyed fewer predators there and didn't realise that they were actually trapped in a 'living larder'.

It is now recognised that it is the biggest fish that produce the healthiest, most vigorous offspring. So the Nhunggabarra method of trapping represents an eco-friendly, sustainable harvesting approach that was able to maintain an abundance of fish.

Further to the south, Mary Gilmore recollected this abundance in her childhood during the 1860s on the Murrumbidgee River. Some of the freshwater lobsters her family caught there were large enough to provide a meal for all four of them. But a few years later they had all been fished out and her father told her that 'when the blacks went, the fish went'.

Gilmore's father once got a surprise when he went to fill his billy at a small waterhole in the 1840s and found a large fish there. It was no accident, either. Aboriginal people deliberately blocked off access to fish from pools in order to keep them trapped in there. When Major Mitchell crossed the Darling River in December 1831, it was probably at another such pool, for he reported 'two large trees had fallen across the stream, from opposite banks ... [and were] interwoven with rubbish'. He had them cleared from his path and white people continue to do this without any inkling of the destruction of the fish resource that will ensue.

For without the original deep permanent pools of southern Australian rivers there are no drought refuges for many aquatic creatures and extinction is hastened for many. Rivers naturally become chains of ponds after long years of drought in Australia. But humans have exacerbated the problem. Rob Gourlay sees the loss of the chains of ponds as a result of the general drying out of the landscape:

'Chain of ponds systems have been destroyed through grazing (trampling), gold mining and land clearing that changed the soil hydrology and dehydrated the landscape. The degradation of the soil health around these ponds eventually impacted on the pond ecology (eg. frogs and other water biology) and thus these living systems collapsed.'

Mitchell once described the Darling River as 'an extraordinary river ... with the finest water ... and without which those regions would be deserts ...' He noted 'water being beautifully transparent, the bottom still visible at great depths, showing large fishes in shoals, floating like birds in mid-air'.

But today we see a different river: a 'turbid stream filled with opaque silt, which allows no fish to breed and no light to penetrate to the plant life' laments author Karl-Eric Sveiby. Since the advent of chemical farming, toxic run off of nitrogen and superphosphate festers in the river, as well as carcinogenic pesticides such as Endosulfan and herbicides like Atrazine, a neurotoxin, that spew from cotton farms. Cotton, the world's thirstiest crop, has spelt death to the Darling.

It's no wonder that the native fish of the Murray-Darling system are down to an estimated 10 percent of pre-European numbers, especially with carp out-competing them and locks and weirs blocking their way. Electronically tagged fish have been found to travel to upwards of 1400 kilometres up the Murray during their life cycle.

On a happier note, the Murray Darling Commission has just finished building five out of thirteen planned fishways (a series of stepped pools beside the artificial barriers) which are being used by most fish species, which have all been found to travel vast distances, it was reported (*The Weekly Times*, 28th March 2007).

Perhaps if Australians could develop a taste for carp (apart from boiling them up as garden fertiliser), as have the Asians, we would get better at reducing this feral pest and saving and restoring our suffering waterways.

Eating Carp
by Janet Millington,
Queensland aquaculturist.

To cook up a nice meal of carp fish you can do pretty much anything you like with them BUT a purging period of 2 -3 days is needed to improve the flavour. The muddy taste of the fish may be caused not by mud, but something in the algae.

What we need to reintroduce is the concept of the medieval English stew pond...where fish from the wild are put in deep ponds close to the house and taken from there for freshness after having been purged (John Taverner described this is 1600s.) If you keep that pond clear and feed fish on barley or vegetable scraps then the flavour should be okay.

Sustainable Aboriginal aquaculture

The clans of the Gunditj Mara, the first peoples of Victoria's south-west plains and coastal areas, lived in the most populated areas in the country. On the lush volcanic plains they built permanent homes and villages that are unique in Australia. One can still see the remains of no fewer than 1000 huts in just one settlement close to Lake Condah, which lies beside the old volcano Mt Eccles, known as *Budj Bim*, meaning 'High Head' in Mara language.

Houses were well constructed, with circular rock foundations and sturdy timber framed domes, some three metres (10 feet) wide and one and a half metres (5 feet) high, all snugly covered with turf. It was all made possible for the Mara by the cyclic presence of water plus their bountiful eel-based economy.

Elaborate stone eel traps captured migrating Short Finned Eels in their fresh water phase. Young elvers spend their early life in the southern waterways, then migrate down to the sea and travel on to their breeding grounds near far-off Vanuatu, on maturity. The Mara trapped them in long woven baskets that would be fixed to stone-walled channels. Artificial reservoirs were also made by the Mara to farm the eels.

The drainage of the Condah Swamp from the 1880s is typical of the damage done to wetlands all over the country. Prior to this, the swamp was abundant with eels, fish and wild fowl, while kangaroos, emus and koalas were numerous in the thick forests around the swamp. The Mara had lived here extremely well.

The newly drained land grew marvellous crops of potatoes in the super-fertile swamp soil. But the soil's fertility was also soon drained away and the ten-acre (4.5 hectare) blocks of the Condah Swamp Village Settlement, given out to unemployed Melbourne families, were too small to make a living from. Most of the early white settlers abandoned their farms, as the Glenelg Heritage Study (2002) explains.

It is possible to view the remains of the Mara culture today, with the indigenous 'Budj-Bim' tour company providing guided tours to the ancient Lake Condah eel and fish traps and the only permanent residences built by indigenous people on the Australian mainland.

On 30th March 2007, after an 11 year battle for their land claim, and with great ceremony, the Gundutj Mara people were finally given back title to some of their land, having successfully proven their strong cultural connections with it. The return of 140,000 hectares of crown land is, so far, only the second such successful land claim in Victoria (and the one hundredth in Australia).

The Gunditj Mara plan to re-flood the lake and restore the ancient fish and eel traps. They also want to gain World Heritage Listing for the unique cultural environment around Lake Condah.

The wealth of swamps

Victoria's swampy wetlands, graced by mighty river red gum trees, were lush larders and an abundance of food and water meant that swamps were also used by Koori women as birthing places too. Testament to much feasting, big middens of food waste used to ring the margins of once fertile swamps on farmland near Hamilton, I was told by a local farmer, who is now keen to revegetate his bare swamps.

Where seasonal swamps have been changed into permanent lakes there have been dire effects. It has been found that swamps are much more important than lakes, in terms of biodiversity.

Northern Victoria's 470 hectare (1000 acre) Big Lake Boort had its outlets raised in 1850. 'This killed every red gum in it' lamented Boort farmer and landcarer Paul Haw. The trend was repeated elsewhere and by the 1950s people were raising flood banks and altering flood patterns as well, he says.

Wetlands continue to suffer from decreasing inflows, salinity, turbidity from sediment run-off, weeds and pests, the loss of riparian vegetation and changes to water flow patterns. Wildlife has also declined and it has been estimated that waterbird numbers in eastern Australia in 2004 were down by 82 percent from 1983 levels (Water Services Association of Australia, via *The Age,* 5th June 2007).

One ex-wetland, Lake Mokoan in northern Victoria, is set to be restored to something like its original. But not if the irrigation farmers of the Broken River Valley get their wish to keep it. The state government plans to close and decommission the lake in 2009, releasing 44,000 megalitres to boost flows in the Murray and Snowy Rivers. Studded by skeletons of the dead trees that once graced it, the irrigators there are hoping for a compromise, with a 'mini-Mokoan' created in the middle, perhaps, to keep their water supplies assured (*The Weekly Times*, 4th April 2007).

Water for the Wimmera-Mallee

North of the Grampians, in the semi-arid Wimmera and Mallee, arable land was magically opened up from the late 1870s with the creation of, ultimately, 17,800 kilometres of hand-dug (with horses) water channels. It was a highly ambitous pipeline project that covered 36 towns and 2.9 million hectares of land. Water was taken from the Grampians, with everyone getting their dams filled during two months of each year. However some 95 percent is estimated to be lost from leakage and evaporation. And in dry times the wind can choke the channels with sand, which then has to be regularly removed.

In the 1990s open channels in the north were replaced with piped Murray River water down to Sea Lake. The underground piping process is slowly continuing. Now wildlife, previously unused to having open water so available, have been studied to assess what the impact on removing this water from them might be, and how this can be reduced. Special drinking stations are to be put in place for those wildlife that have become reliant on the channel water's availability.

References:

Morrison, Edgar, *'The Loddon Aborigines'*, self published, 1971.

Keating, Jenny, *'The Drought Walked Through'*, Department of Water Resources, Victoria, 1992.

Cunningham, Irene, *'The Land of Flowers'*, Otford Press, Australia, 2005.

Bayly, Ian, *'Rock of Ages'*, University of Western Australia, 1999.

Hedditch, K, *'Land and Power-* A Settlement History of Glenelg Shire to 1890'. Glenelg Shire Council.

Fairly, Alan, *'A Field Guide to the National Parks of Victoria'* Rigby, Australia, 1982.

Sveiby, Karl-Eric and Skuthorpe, Tex, *'Treading Lightly'*, Allen and Unwin, Australia, 2006.

Glenelg Shire Heritage Study, Victoria, 2002.

Aboriginal gardens of the Murray River and its tributaries: - www.cbonline.org.au/index.cfm?pageId=36,87,23,57

Budj-Bim Tours (of south-west Victoria's Gunditj Mara heritage), website -www.wmac.org.au

1.3 Ground Water

Ground water origins

Conventionally speaking, groundwater is a stage in the hydrological cycle, whereby the downward percolation of rainfall is stored underground in aquifers, the upper portion being known as the water table. This water often then flows up to the surface as springs and contributes to river systems and other surface waters.

Groundwater from bores and hand dug wells has historically been only a minor, although important, source of supply in Australia. Things change when rain is scarce and surface waters dry up. Groundwater starts to become an attractive option and there is usually a mad rush for drillers and long waits for licences to gain permission to sink the bores (with the number of applications in 2006 at up to three times normal levels).

Years of drought in southern Australia have resulted in reduced aquifer recharge; however, there is 'no single understanding or definition of sustainable yield' of this groundwater, laments the Centre for Groundwater Studies.

So where does this water originally come from and how can the sheer volume of water in Earth's oceans be explained? It surely can't be explained purely as the condensation of water vapour in the atmosphere. Australian scientist Rob Gourlay points out that:
'ocean waters have increased and risen over the past 17,000 years by about 79 metres and the source of this water must have been from deep within the Earth, as the ice melts could not account for this volume.
'... Scientists probing the Earth's interior have found a large reservoir of water equal to the volume of the Arctic Ocean beneath eastern Asia. This seismic anomaly is under the Asia plate and within the mantle at a depth of roughly 1000 kilometres (620 miles). Other such groundwater reservoirs exist within the Earth's crust and mantle'.

While there is evidence for water coming down from outer space in comets, other scientists have been looking deep beneath the Earth's surface and made some fascinating discoveries there.

Since the late 1960s Earth scientists have had technology available to allow them to look deep within the Earth. This has been useful for them in understanding earthquakes and volcanoes.

A geochemist from Scripps Institute of Oceanography in California, Harmon Craig, discovered helium-3 around 1970. This is a rare isotope formed during the birth of the solar system and most of it is locked deep in the Earth's interior. Quantities are also found along the boundaries of continental plates, such as the Pacific Ocean's 'ring of fire' and around volcanic 'hot spots' on ocean floors. Craig found the greatest of concentrations about 48 kilometres (30 miles) off the coast of Hawaii, above a 13 kilometre (8 mile) deep volcanic vent.

On a 915 metre (3000 feet) dive there in a high tech submarine with colleague David Hilton they saw a huge glowing crack in the sea floor surrounded by 'rust'-covered terraces of basaltic lava. He said, in *The Water Dowsers Manual,* that:
'Shimmering, champagne-like water bubbled up out of the vent – water hot enough to rise but so stuffed with carbon dioxide, according to Hilton, that it actually flowed down.'

The Russians have stumbled on deep underground water too, in their super deep drilling projects since 1970, the American Society of Dowsers also reports. The geologists were amazed that at around 4575 metres (15,000 feet) they have found what geologist Y A Kozlovsky called 'surprising copious flows of hot, highly mineralised water.' According to the text books, this was impossible!

There is a reference in the Bible to 'the fountains of the deep' opening up. Could this be talking about similar deep waters that have never before seen the light of day? The sort of water that flows under pressure from under the ground in defined geological pathways and which is located the world over by water diviners?

Stephen Reiss was a mining engineer who was also confronted with inexplicable and abundant waters of the deep that often spoiled mining operations. He was not the first to notice them, Christopher Bird reported in his classic book *Divining* – a 'bible' of dowsing. Reiss had said in 1954 that:

'For a century it has been known that, under certain conditions, some rocks yield hydrogen and oxygen gases which subsequently combine to form 'new water'. In connection with the mining and recovery of gold, a natural coincidence led me to suspect, many years ago, that such a laboratory reaction might proceed within the Earth.

'My discovery was then put to a field test by locating and drilling many water wells. The record to date is 70 producing wells out of 72 attempts, all drilled in hard rock, all located in distress areas generally considered unproductive'.

Classical authors concur with Reiss. Plato and Aristotle believed that water was formed within the Earth, and around 500BCE Anaxagoras said that oceans were created from rivers flowing into them as well as from the 'waters of the Earth'. Vitruvius, in Roman times, urged people to look for the best water supplies not in sands, gravels and soil – but in rocks. Seneca and Pliny the Elder also wrote of underground rivers flowing in veins that 'pervaded the whole Earth within and ran in all directions bursting out even on the highest ridges' (and where hydrologists wouldn't bother to look for them).

Deep water supplies have provided sustained volumes of fresh water to many urban settlements, in often arid parts of the world, over vast time spans. Sources which just cannot be explained by conventional hydrology. Beirut gets its water supply from bores in the mountains of Lebanon; Damascus, the oldest continually inhabited city on Earth from 7000 years ago, has its ancient Ain Figeh spring gushing out from under a rocky mountain.

Reiss went on to find many more often phenomenal water flows from the deep and provided many reliable supplies for American communities over the subsequent decades. Others came before him with modern understandings of new water, writes Rob Gourlay:

'Primary water was postulated by Adolf Nordenskiold in the nineteenth century, and raised in the book "A Journey to the Earth's Interior" by M.B. Gardner (1913). Nordenskiold wrote an essay on the subject of primary water, which resulted in him being nominated for the Nobel Prize in physics.'

In 1960 Michael Salzman, an engineer and professor at the University of California, published his book 'New Waters for a Thirsty World', which received much flack from the hydrologists at the State Department of Water Resources. Reiss has now passed on, but Morad Eghbal and the Reiss Institute continue his research work. Eghbal writes that:

'Conventional hydrology speaks of a static supply of water created once early in the Earth's history and being constantly recycled .... Stephan Riess saw new additions of water flowing vertically, from beneath the surface and adding to the hydrologic cycle. This water in turn, becomes bound up on the surface partially in plants, sediments and subduction zones on its way back to the Earth's mantle. These new additions occur frequently where there is faulted, igneous and metamorphic rock, and they can be intercepted to replace contaminated supplies and provide new sources of water in arid areas.'

Rob Gourlay explains further that:

'Primary water is generated in the rock strata when the right temperature and pressure is present. This water is then forced into fractures and fissures in the rock where it can transverse large distances, hundreds of kilometres. Some of this water is expressed as springs, which can be either hot (thermal) or cool. This water is always moving and therefore can be detected by dowsing.'

Reiss believed that the new water is formed as a result of the outgassing of volcanoes deep under the Earth's surface. These eruptions create gases that either escape to the surface as gas or turn into primary water.

Testing of the dissolved gases coming up with water found by the Reiss Institute has revealed increasing concentrations of argon, CO_2, methane, ethane, helium and tritium the deeper down the waters are sourced from. Clearly these signature components indicate a source other than rainfall.

New water may also be constantly manufactured in deep rock strata in association with the daily flexing of the Earth's mantle, surmises American dowser John R McCreary (in *The American Dowser*, November 1981):

'Intuitively, I would like to suggest that this manufacture comes about through the dielectric build-up of an electrostatic charge from the daily stress of the Earth's crust as the planet rotates in relation to the

sun and moon. Because it occurs so gradually, we are not physically aware of this daily average rise of 11 inches (28 centimetres), which peaks about 80 minutes after the sun is overhead. It produces sufficient charge by stressing rocks containing quartz or similar crystals to permit the interchange of the electrons and protons with the subsequent bonding of the nucleus to those of the hydrogen. In my opinion, this daily flexing phenomena aids and abets the penetration and flow in all of these complementary water systems'.

Groundwater knowledge in Australia
by Rob Gourlay

'There is a public and government perception that groundwater is a limited, unsustainable, non-renewable, contaminated and untouchable resource. This is why Governments have not developed mapping technologies and exploration arrangements that facilitate groundwater assessment in a manner similar to minerals exploration

The skill in Australia for finding deep, fractured rock groundwater largely lies outside of the public sector and this explains the deficiencies in government water policy and the mapping of deep groundwater locations and groundwater availability.

The public science position, that groundwater exploration is unsustainable, other than in Western Australia, is simply nonsense. There are massive groundwater resources throughout Australia, including the fractured rock aquifers and buried palaeo-channels [old rivers] in desert areas. The deserts of Egypt, North and South America are renowned for these deep, buried drainage systems. Also, a groundwater bore at Gumly Gumly near Wagga Wagga, in New South Wales, produces a water flow that services a rural community greater in area than Tasmania.

There is also a deep palaeochannel (100 metres underground) west of Forbes that was once the old Lachlan River. It once flowed in a south-north direction and it now flows under the current Lachlan River, without any physical connection due to the thick clay layers (aquiclude) between the old and the new rivers.

Cherry farmers in Young draw groundwater from 300 metres down through solid granite rock and there is little or no physical connection between these groundwater extractions and the sustainability of local rivers or other surface water resources.

I also think that organisms may be involved with electrical and chemi-

> cal reactions (remembering that marine organisms have been around for about 3 billion years or more and organisms within the rock system may have been around for much longer). ... And since water can be created by organisms (eg. converting hydrogen peroxide H_2O_2 in the soil into water, by stripping away an oxygen molecule) the amount of water on the Earth is potentially increasing; with some more water coming in from outside the atmosphere ...
>
> Lack of knowledge of the soil water processes and groundwater water availability has led to claims by Clive Hamilton (opinion writer and ex bureaucrat) and the Wentworth Group that farmers should be removed from the so-called marginal farming land.
>
> Admittedly, conventional farming practices with ploughing and chemical fertilisers do degrade soil health and dehydrate the soil, however, removing farmers fails to recognise the environmental impacts of this action. In this situation, there would be new public costs for controlling woody weed invasion, soil erosion and fire management (global warming exacerbates these issues). Farmers in these areas have been failed by public science that did not understand the opportunities to build soil health (by restoring soil carbon, nutrients and microbes), increase soil water storage and utilise our significant groundwater resources to buffer drought periods.'

How do they find New Water?

How does the Riess Institute know where to drill for new water? A brochure states that they:

' ... *use mineralogy, petrology and structural geology precisely to locate high pressure/low temperature hydrothermal systems that have previously been encountered randomly by engineers in mine and tunnel flooding incidents.*

'Recently, with sophisticated satellite photography and "remote sensing", water can be found in rock using a technique called "fracture trace analysis". Large fractures are identified by satellite photography for exploratory drilling.'

And how does Rob Gourlay determine where to look for supplies of new water for clients in Australia? Gourlay combines a high-tech approach with good old fashioned water divining. In October 2005 he said that he uses:

'... geophysical data to target good fractured rock zones and this has paid-off for me over the past 6 years with over 300 bores. The primary water I locate is normally of very good quality and this is why I target deep fractured rock water. ... A problem that I have encountered (once) with fractured rock is high iron, however this was not that deep at 90 metres.'

So does Sydney really need de-salination plants to solve its water crisis?, I asked him.

'Sydney has a massive buried stream network system, about 100 metres down, that flows from the west to the ocean ... I believe that most factories and other large water users should tap into the deep fractured rock groundwater system to supplement other water sources. All the more reason to relocate large water users out of cities (where shallow groundwater can be contaminated) into major groundwater zones in regional areas.

'Groundwater as fractured rock water, is potentially the greatest source of sustainable water supply ... [However] the shallow groundwater that is rainfall fed, along with surface supplies from rivers and dams etc, are at the greatest threat of over-exploitation.'

Photo: Rob Gourlay

Australia's Great Artesian Basin

Definition of artesian
Artesian Basin – a geological feature in which water is confined under pressure. Artesian bore – a bore going into an aquifer in which the water level rises, by hydrostatic pressure, into the shaft to above ground level. Origins – from 'Artesien' the gushing well waters of the former French province of Artois. (*Macquarie Dictionary*)

The massive Great Artesian Basin covers around 25 percent of the continent, at around 1.7 million square kilometres – making it one of the largest freshwater basins in the world. The Basin is thought to be recharged from a zone of porous layers of sandstone that lies on the western slopes of the Great Dividing Range, spanning from far north Cape York in Queensland down to New South Wales. Rainwater percolates downwards in the highlands and then flows very slowly west and southwards to the south-east corner of the Northern Territory and north-eastern South Australia.

There has been an old theory that some of the water in the Basin also originates from New Guinea. It is not such a far-fetched idea, when you look at a map of the Basin. You'll see that the Basin extends right up to the top of Cape York Peninsula in Queensland and it's easy to imagine that it goes further. Both countries were once connected until sea levels rose.

Bores sunk into the Great Artesian Basin run like a tap, without the need for pumping. But a century of uncontrolled use and wastage of these waters has led to a big drop in water pressure. A program of capping and piping of open bores is under way, but in 2003 892 bores remained uncontrolled and flowing freely.

Desert oases

It's fairly arid above ground in the Basin region, but in some areas there are oases of abundant life, in the form of mound springs. Unique eco-systems with endemic species only found in the Basin, these special springs are also drying up. They are found along the Great Artesian Basin's south-western edge, in an arc of about 400 kilometres, where the Basin's water reaches the surface through fault lines in Queensland and South Australia. The mounds, often as big as small hills, form from the sediments and salts brought up and deposited by the naturally arising artesian water as it evaporates over long periods.

The mound springs are important cultural places for the Arabunna people, the traditional custodians of the area. But the fragile eco-systems are under threat. Winston Ponder of the Australian Museum laments the demise of the mound springs:
  'Many springs have disappeared in the last 100 years as a result

of water extraction from the Great Artesian Basin, probably resulting in the extinction of unique species before they were even discovered.

[They are]'... home to a diverse array of unique and unusual aquatic invertebrates and fishes. Furthermore, they are also important habitats for plants, birds and other terrestrial animals in an otherwise barren landscape'.

No doubt the mound springs around South Australia's great water guzzling Roxby Downs/Olympic Dam and Beverley uranium mines are some of the most profoundly affected. There are about 600 individual springs, in 11 major groups, across the Great Artesian Basin. The southwestern region of the basin, which is closest to Roxby Downs, contains the largest number of active, unique and fragile springs.

Currently BHP Billiton extracts 33 million litres a day from the Great Artesian Basin. Farmers, environmentalists and traditional owners have reported dramatic reductions in water pressure threatening and sometimes extinguishing the rare ecosystems. The Olympic Dam mine is the single biggest industrial user of underground water in the southern hemisphere.

Under the *The Roxby Downs (Indenture Ratification) Act 1982*, BHP Billiton is not required to pay for this water and its water use is exempt from the Water Resources Act 1997. On 2nd February 2007 Prime Minister John Howard made a statement suggesting that perhaps BHP should be paying for this water.

Now the mine wants to expand to become the world's largest uranium mine. BHP Billiton has put a proposal to the governments that would see the company extract an additional 120 million litres of publicly-owned artesian water per day, every day, for the next 70 years!

The Friends of the Earth have called for pressure to be applied to stop the expansion and said, in May 2006:

'With BHP Billiton seeking a four-fold expansion of their Roxby Downs mine and the Indenture Act due to come up for review in the next 18 months, now is a crucial time to act for government and corporate accountability.'

Unleashing the unthinkable?

Back in the early 1980s I worked with Greenpeace, Sydney, in trying to stop the Olympic Dam uranium mine from going ahead. A blockade attracted 700 dedicated people from all corners of the country to camp beside the proposed mine site. It was a huge effort to get them organised, to train them in non-violent action in advance, to provide food and shelter etc.

The country was beautiful to behold. Not a desert as some would imagine. It was vibrant, vegetated land of fat Sleepy Lizards, diverse trees and shrubs and Sturts Desert Pea flowers like red jewels on the sand. We knew the importance and fragility of the mound springs; the unspeakable dangers of what humans can do with uranium. A huge campaign to prevent this mine had been passionately fought, including two blockades. But it went ahead.

The mine shaft was actually sunk exactly over the crossing of a couple of Dreaming tracks, a site sacred to the Kokatha people. Further north, a legend from the Ranger uranium mine site, in the region of world heritage listed Kakadu National Park in the Northern Territory, warns that dreadful gigantic ants live under the ground there and, if disturbed – they could emerge and destroy the world!

Twenty five years have gone by since my direct protest days. For a long time uranium prices were not high enough for the huge multinational miners to get excited and some mining leases were not mined. Jabiluka was supposedly mothballed to appease Aboriginal land owners. The catastrophic nuclear accident at Chernobyl was a generation ago now (in April 1986) and the memory of such disasters has grown dim.

Now we have a new era, with the scenario of climate change. Uranium's image has been re-forged as 'green', because nuclear power production has low greenhouse gas emissions – therefore it must be good in an economy based on carbon emissions! Suddenly the Labor Party has thrown out its 24-year-old policy of only allowing three uranium mines in Australia. Perversely, it can even consider itself green with this new direction (although there's still opposition from many party members).

Is the Basin running dry?

In November 2005 Professor Lance Endersbee, a former Monash Engineering dean and pro vice-chancellor, sounded alarm bells in his book, *'Voyage of Discovery'*. He warned that the Great Artesian waters were not recharging, were virtually fossil waters of the Earth, an unrenewable resource running dry.

If correct, and water extraction from the Great Artesian Basin has been completely unsustainable and it will run out, it is terrible news for the Australian economy and life in the outback. But John Hillier, of the Great Artesian Basin Consultative Committee, believes he is wrong about the recharging of the basin:
'As more knowledge on water flow directions, water quality and the geological structure has been gained, it has become obvious that the Great Artesian Basin performs like any other groundwater basin, except that it is much larger'.

The GABCC's Jim Kellett explains that:
'We have irrefutable evidence that the Basin works as we understand it to. We've done $Carbon_{14}$ and $Chlorine_{36}$ isotope dating on the water extracted from bores across the basin. These data clearly show an age gradient, with the oldest water being in the centre of the feature, and the youngest being close to the intake zones in the highlands .... The oxygen and deuterium isotopes have the distinctive signature of rainfall.'

I'd say Professor Endersbee is partly correct, as primary waters may well be supplementing rainfall inflows to the Basin and helping to pressurise the system. In the end summary of his book he states that 'the volume of water at the surface of the Earth is increasing very slowly'.

Getting into hot water

Most geothermal heat in the Earth's crust is from the radioactive decay of elements such as uranium, potassium, and thorium, the Department of Primary Industries informs us. Granite is naturally radioactive, containing uranium and similar elements, while radioactive radon gas is outgassed from it. The DPI explains that:
'Most of Australia's hydrothermal resources are water-dominated

sources found in the Great Artesian Basin ... For a long time, the temperature of the bore water was seen as a nuisance because it had to be cooled before stock could drink it'.

In other parts of the world geothermal heat has been regarded as a boon and well utilised. The majority of the population of Iceland, for example, live in homes heated by hydrothermal energy. But there has been a little energy harnessing done here, the DPI reveals:

'The first hydrothermal electricity installation in Australia went into operation in May 1986 on Mulka Station in the remote north-east of South Australia. A Rankine Cycle engine (using a freon refrigerant) was attached to the station's hot (85 degrees Celsius) artesian bore to generate 20 kilowatts of continuous power. This was the lowest temperature hydrothermal resource which had been successfully used anywhere in the world. A 150 kilowatt unit has also been operating at Birdsville (in Queensland) from the town's 99 degrees Celsius bore'.

Eminent scientist Tim Flannery was flagging geothermal as Australia's best potential energy resource, in an interview on ABC Radio on 5th February 2007, saying that:

'There's one geothermal area in northern South Australia that is capable of providing ALL of Australia's energy needs for the next hundred years. It would take about ten years to develop and would cost about the same as coal fired power generation; while it is a renewable supply, as the underlying granite breaks down continually fuelling the heat. You just need a heat exchanger to convert it to power'.

Hot springs in Australasia

Hot mineral springs are commonplace in volcanic places like Iceland, Japan, Bali, New Zealand and Polynesia. In such places they have often played a highly significant role in cultural traditions.

In Australia it is quite the opposite situation and very little use has been made, although a few geothermal bathing pools can be found, thinly scattered far and wide. But they are often worth the distance travelled!

Alanna Moore

Geothermal New South Wales

Artesian waters of Lightning Ridge

Lightning Ridge is a one day drive from Sydney, in far north central New South Wales. Famous for its black opal, which is found fairly close to the surface by rugged individuals, it's a colourful town of miners, tourists, artists and Aboriginal people. Winter is the peak tourist season, when the weather is mild and pleasant. What is not so well known is the presence of an outdoor hot artesian bathing pool, which is open to the public 24 hours a day, seven days a week. And it's free!

Hot, mineral rich waters gush up from over 1000 metres below the surface and are mixed with a little cold to keep a constant temperature of 40–42 degrees Celsius. This is then channelled into two circular concrete bathing pools. There's a large swimming pool where weary opal miners swim, float, lounge, relax and socialise with neighbours after a hard day's mining, as well as a shallow childrens pool.

People from all over are attracted here to enjoy the Bore Baths, especially those from the colder states of Australia during the winter months. They come seeking the therapeutic power of the water, which is rich in potassium salts, and is particularly good for ailments of a rheumatic and arthritic nature. The regular long pilgrimages of some visitors must bear testimony to the value of the pool. Such special waters are highly appreciated back in Europe and the Bore Baths are often full of European bathers.

In the past 20 litres of water per second flowed up under natural pressure and continued on through 100 kilometres of open drains to supply stock water to grazing properties. An estimated 450 million litres a year were wasted through evaporation and seepage. In 1997 the *Cap and Pipe the Bores* program plugged the bore and laid 170 kilometres of underground pipes to replace the open drains. The bore supplying the Bore Baths was capped to reduce the flow to 9 litres per second and the used water from the bathing facility is now sent to the opal fields for 'puddling' (washing the mined rock to look for opals).

This has helped preserve water availability and sustainability. New South Wales has around 1400 artesian bores of which half have stopped flowing due to huge wastage and pressure reduction. There are some 7000 kilometres of open bore drains in the state where up to 95 percent of the water is lost through evaporation, seepage and breakouts. These are all gradually being replaced by underground pipes.

South-east of Lightning Ridge is Moree, with a population 10,000, it is located 628 kilometres north-west of Sydney. Billed as Australia's 'artesian spa capital', the town's hot artesian spa bath complex has evolved from the town bore, which was originally sunk to a depth of 850 metres in 1895. Baths were soon set up (different days originally being allocated to the different sexes) and the local council began promoting the waters as curative and a source of relaxation and replenishment – a tradition still carried on today. The bore ceased to flow naturally in 1957 and is now worked by a pump. The mineral-rich water emerges at 41 degrees Celsius and is pumped into the public pool complex via both underwater spa jets and above-pool spouts. The pools are emptied and cleaned each night.

The 'Hot Springs Health Resort', on the corner of the Newell Highway and Jones Avenue, Moree, has three geothermal pools. The proprietors state that the waters 'help ease stress, restore nerve and muscular functions, relieve inflammations and painful joints, support acid-alkaline balance and strengthen the immune system.' Water comes up from nearly one kilometre below and is also sold in bottles in Moree's supermarkets.

In Walgett, not far from Lightning Ridge, there is also a hot bore bath and swimming pool – but both are currently closed at the time of writing, due to the severe drought. Between Walgett and Wee Waa is the Burren Junction Bore Bath, an outdoor, open facility that's free to use, with a pleasant camping area adjacent. This was also closed for a facility upgrade, which includes capping the bore.

Walgett Shire's website states that:
'Burren Junction Bore Bath will be upgraded to include filtration and disinfection systems, and will be accessible to wheelchair bound patrons (subject to medical approval and with assistance). The pool will include seating for up to 20 people and will be covered to prevent heat loss. The pool will comply with all relevant design and health guidelines.'

As for the Lightning Ridge Bore Baths, the Shire tries to cover any liability with a sign which warns that the 'water does not comply with current standards. Patrons are advised that they use the facility at their own risk.' But given a choice, I'd personally prefer to enjoy the Bore

Baths at Lightning Ridge. They are simple and close to nature, while kept clean and hygenic enough. There is no roof, no fences and bathers are free to come and go as they please. Who could resist lying in the steamy hot water in the shade of adjacent trees in the early morning light, before any crowds come. A peaceful spot where you might watch the wildlife or wildflowers around you as you soak, occasionally hopping out for a spell to cool down, before returning for more. Or soaking in the soothing waters under a mantle of the stars at night, safe from winter's chill. Pure magic!

The artesian bore also provides enough for all of the town's reticulated water supply too. Can it really all just be explained as purely rainwater from Queensland, seeping south through the Basin at a very slow rate? New South Wales government literature gives us a hint of possibly more primary origins:

'Lightning Ridge lies in a large geological feature called the Surat Basin, which is part of the vast Great Australian Basin ... Water samples collected from deep artesian bores in the New South Wales section of the basin have contained methane gas accompanied by varying amounts of carbon dioxide. Most samples record small quantities of ethane and a few bores contain traces of propane and butane. The Department of Mineral Resources is conducting ongoing studies to determine the source of the gas in the artesian water.'

Geothermal Victoria

Amidst the ancient sand dunes, the tea treed sandy humps and hollows that characterise the Mornington Peninsula, south of Melbourne, there is a brand new geothermal spa centre. 'Peninsula Hot Springs' is Victoria's only naturally hot mineral spring bathing facility. Located in the Cups area of Mornington Peninsula, it has been open since June 2005. The Peninsula itself is a volcanic landscape, with a geological fault – the Selwyn Fault – running down along it.

At Peninsula Hot Springs they use two sources of water, with one, taken from ten metres under the ground, providing a bountiful supply of pure (and deemed potable), cold water which the company has a permit to draw upon for irrigation and other needs. A lower geothermal aquifer provides the hot water that is used for the bathing facilities and this is pumped up from 637 metres below. With its own pressure it comes up to within 10 metres of the surface. Here it is scalding hot at over 50 degrees Celsius and must be mixed with the cold water for bathing.

Over to the west of the state, in the town of Portland, hot water (of around 58 degrees Celsius) has been extracted from four bores 1400 metres deep at a rate of 65 litres a second since 1983. This is used to heat more than 19,000 square metres of buildings by underground reticulation.

It is Australia's only geothermal space heating facility, providing an estimated 3.6 megawatts of heating to local public buildings as well as heating the municipal pool. After the heat has been extracted from it, the water goes to the cooling tower, where it cools down some more and any H_2S gas and iron is removed. From there it goes into the city's water reticulation system for consumption.

Taking the spa

In various parts of Europe patronage of 'spa bathing' facilities and the drinking of special mineral waters has been popular since prehistoric times. English fascination with healing waters had waned, but was revived from the sixteenth century onwards. Many people caught the craze and the well-to-do took to visiting Spa (or Spaw, in Belgium) and other famous mineral water centres on the Continent. They then applied the name generically to other such facilities back home, so that 'spa waters' today can be taken in many places.

Spa bathing continues to provide relief to many sufferers of pain and illness, who find deep relaxation and healing this way. Gushing warm waters that pound your body stimulate the skin, helping the elimination of toxins and causing endorphins – nature's own pain killers – to be released. (Endorphins also kick start internal body regulators.) Blood flow is enhanced in a hot mineral bath, muscles relax, sinuses may clear and the body heals itself quicker, it has been documented. Hydrotherapy is the new buzzword for this ancient healing modality.

Australian mineral waters

According to the legal definition in Australia's *Ground Water Mineral Water Act of 1980*, mineral water must in its natural state contain carbon dioxide and other soluble matters in sufficient quantities to have effervescence and a distinctive taste.

Often this water is marketed as 'spring' water. The definition of what constitutes a spring is somewhat lacking however, while it adds a romantic and authentic ring to the often lacklustre waters that come in bottles. In reality the bottled water business is not at all romantic. Waters are more likely sourced from a bore, or possibly even from a mains supply tap. And minerals are often artificially added.

Companies like Coca Cola Amatil have a bad trackrecord in over-extracting ground water for bottling in places like Peats Ridge and Gosford, on central coast New South Wales, and Mt Tamborine, in south-east Queensland. Their 'Peats Ridge Springs' water, when tested, turned out to be the most highly acidic and to have the highest aluminium content (200 times that of Sydney tap water) of all brands tested, it was reported in November 2004, writes author John Archer. The bottled water industry remains under-unregulated.

Some natural spring water is high in sodium chloride/salt, which can harden arteries if taken in excess. Calcium and magnesium salts in mineral water determine its level of hardness. While hard water is a nuisance for washing, will not lather easily and produces scale in jugs and plumbing, there is a good side to it. There are fewer rates of cardiovascular disease where water is hard.

'Spa Country' – the Central Highlands of Victoria

At Mordialloc back in 1856 artesian water was, for the first time, successfully tapped into in Victoria. What became of it is not known, but certainly the mineral springs of Victoria's Central Highlands regions to the north west developed into the most numerous and well known in the country.

Aborigines would have already known all about them, of course, and no doubt relied upon the Central Highlands springs during the dry months of summer. A member of the Jajawurrung tribe is said to have guided the early Europeans to the springs (which is generally how all the earliest water supplies were 'discovered' by white people in Australia). They were only reported from 1836, after Captain Hepburn 'discovered' the Hepburn Mineral Spring. By the 1850s the region was in the midst of a gold rush and the natural environment was being devastated. Creeks were mercilessly mined, the forests were felled and land became cratered with diggings. Many more mineral springs were unearthed at this time as a result of mining diggings.

Fortunately the forests have since regenerated and, with the pleasant terrain, quaint cottages with their English-style gardens, a mild climate in summer and good spa facilities, it has become a popular area for city visitors for many decades, being only around a one hour drive from Melbourne. Daylesford and Hepburn Springs now enjoy tourism as their main industry, with the lure of soothing and therapeutic waters for luxury spa bathing, plus massage and other therapies and lots of gourmet food – it's the pampering capital! After ten years of drought, current water restrictions in the region disallow the filling of spa baths. But tourist resorts are exempt!

Further east in the Central Highlands area, Kyneton boasts one of the last few remaining mineral water bottling plants. Kyneton is a small town of impressive 19th-century bluestone buildings, set on a fertile basalt plain 509 metres above sealevel. It is the most accessible springs area to Melbourne, which is 88 kilometres away, down the Calder Freeway.

Where does the mineral water come from?

The area of central Victorian where mineral springs are found is about 18 by 60 kilometres and around 65 springs have been recorded there. The area has had much recent volcanic activity, with about 36 extinct volcanic cones dotted around the Hepburn Shire alone.

The definitive book on the Spa Country springs, authored by Edward (a scientist and glacier expert, retired) and Maura Wishart, tells us that:
   'The majority of these springs discharge directly from Ordovi-

cian layers of rock, usually sandstone, shale and silt, which were laid down under the sea about 450 million years ago. In central Victoria the aquifer consists of folded, faulted and fractured Ordovician sedimentary rocks which extend from the water table to a depth of at least 1000 metres in some places.

'...*In the mid 1970s, bores sunk to a depth of 80 metres by the Department of Minerals and Energy intercepted mineral water. During underground gold mining activities in the early 1900s mineral water was struck at a depth of 255 metres.*'

What's in the mineral water?

Typically, central Victorian mineral waters will contain up to 2.5 grams of total dissolved mineral salts per litre. The minerals present, generally in the user-friendly form of bicarbonates, include calcium, magnesium, sodium, potassium and a little iron bicarbonates, plus mineral compounds of chlorides, sulphates and silicates as well as other trace elements.

As for Peninsula Hot Springs water, it 'tastes like a highly mineralised sports drink and has a slight sulphurous smell that the team refers to as "real",' says the blurb, explaining further that:

'The geothermal water contains a total of 3700 parts per million of total dissolved minerals. The primary mineral groups include bicarbonate, magnesium, potassium and sodium. Good levels of calcium, boron, selenium and several key trace minerals are also present in the water.'

Bubbly waters

If you go down to the forested valley floor of the Central Springs area of Daylesford, you will be able to see several mineral springs located along the Wombat Creek, all adding to its flow and giving it interesting characteristics.

If you sit awhile on a bench by the Sutton Spring alongside the Creek, you'll probably find it a very pleasant spot. Should the sun be shining and the bees buzzing loudly in wattle flowers above you, you may get pretty relaxed there! (More about that aspect later!) Stare into the waters of the creek and you'll see little groups of gas bubbles bursting to the surface here and there, more or less continuously.

In the past, at another place where gas bubbles were seen – in the Campaspe River at Kyneton – several bores were put down at spots close by. Drillers had to go down through a hard layer of basaltic lava, sometimes up to 37 metres thick, before they reached the mineral water. Hand pumps, for public access to the water, are now installed, and can be found in the riverside park opposite the bottling plant to the north of the town.

Storm fronts are preceeded by a lowering of the air pressure and increased effervescence of mineral spring waters occurs before a storm. More gas is released with the water at these times. When air pressure is high on calm sunny days, the same spring waters are much more still, say the Wisharts.

It is the carbon dioxide in these waters that makes it 'soda water'. This weak carbonic acid gives the water the ability to dissolve the mineral carbonates in the Ordovician rock and form the mineral bicarbonates found in the water. Drinking a lot of it could lead to mineral imbalances, a naturopath warned me long ago. However, it can also be a very healing experience to bathe in the mineral spas. Not only are the minerals in the water taken up by the body, but even the carbon dioxide can have a good effect, studies have shown. Some CO_2 is absorbed via the skin and can help to dilate blood vessels, aiding circulation. CO_2 also inhibits bacterial growth in the water.

What you may not want to drink

Springs with high sulphur content have a characteristic rotten egg gas smell. In fact, you rarely see people actually drinking these waters. Certainly not more than a polite (or impolite!) mouthful. And there is another factor, not commonly known, that can be a real turn-off these days, although it may not necessarily cause a problem.

Much of Victoria is a vast volcanic plain that stretches from near Bendigo, in the north central area, down to Melbourne, across to the Western Districts, and continuing over the border into South Australia. This is actually the third largest volcanic plain in the world. Around 500 extinct volcanoes are found here and Aboriginal eruption stories tell us that they were still eruptive in human memory, perhaps as late as seven thousand years ago, Neville Rosenberg, a local geomorphologist informs me. The youngest volcano – Mt Schank, near Mt Gambier, across the border in South Australia – has been dated to about 4000 years of age. The Wisharts mention 50 volcanoes in the mineral springs districts, Wombat Hill and Mt Franklin being well known examples. (Yes, that sacred ex-volcano Mt Franklin, the 'Uluru of central Victoria', is the source of the name of the well-known bottled water brand and no, it is not bottled there.)

The carbon dioxide in the region's mineral waters is 'thought to be a product of vulcanism' they say. And there's more evidence for this, particularly around the Lyonville/Bullarto/Musk area, where there are seven extinct volcanoes.

I have tuned in to the Earth energies at this point (while dowsing at a vibrant organic farm there) and they are very powerful indeed! I was at the location where the headwaters of the Loddon River rise up, in a pleasant little glade, from the depths below, and go on to wind their way eventually to the Murray River.

Spring waters here have a significant distinction. They are radioactive! They carry radon gas with them, although this outgasses quickly and levels of radioactivity are only very slight.

It is thought that one massive magma chamber fed those seven volcanoes, producing much carbon dioxide, radium$_{226}$, and the gas radon$_{222}$. (In the gold rush times a miner was once 'suffocated by carbon dioxide' while mining underground at Sailor's Falls, south of Daylesford.) When the deep down molten lava occasionally spews up from such hot spots, it often makes its way up to the surface through geological faults, which is why many volcanoes are often found in linear rows.

This one large hot spot was possibly the source of all the carbon dioxide in the region's springs. The Earth's crust, slowly moving on its tectonic plate, passed over the magma vent, resulting in volcanic activity. That hot spot is now currently located beyond the south east coast of South Australia, Rosenberg told me.

The mineral springs of Lyonville have the highest levels of radioactivity in the Central Highlands and the further you get from them, the lower is the dose. The Wisharts explain:
   'The concentration of radon gas in the majority of the mineral springs

of the Central Highlands is around 75 Becquerels per litre. The exceptions are Leitchs Creek with 1443 and Lyonville township with 6600 Becquerels per litre.'

Fortunately radon gas, which is odourless and colourless, is quickly dispersed. It is only a problem if inhaled and has a half life of less than four days. In America radon gas inhalation has been cited as a major cause of lung cancer. In Cornwall homes built in granite country have 100 times the radon gas levels of the British national average and need to be well ventilated; while lung cancer has been a hazard for uranium miners, the *Sydney Morning Herald* reported in 18th July 1985.

Radioactivity used to be considered therapeutic in times past, when people flocked to take it in by frequenting certain caves and drinking particular spring waters. The famous hot springs at Bath in England and King Haakon's in Norway are slightly radioactive. People flocked to the central Victorian springs to take the therapeutic waters, too, especially between the two world wars.

Perhaps a homeopathic dose of radioactivity really is good for you. This was the surprising conclusion made by researchers on a television documentary about Chernobyl 20 years after the nuclear power plant accident. Certainly homeopaths have made good use of radioactive materials as remedies.

Shaun Ogbourne described visits by him and other dowsers to British radioactive springs at the British Society of Dowsers International Congress of 2003. 'When you spend a bit of time around them you start to feel very sleepy' he told us. Perhaps radon is a relaxant!

In Lyonville the Radio Springs Hotel used to publicise itself as being 'only 15 minutes walk to radioactive springs with the greatest radium content throughout the springs district.' And Daylesford sold itself, in a 1928 Tourist Bureau advertisement, as being 'Famous for Radioactive Mineral Springs'.

Hiroshima and the A-bomb put an end to radium tourism. Since the 1950s nobody has mentioned the fact! But none of the springs are found to have unhealthy amounts of radioactivity to worry about. (More of concern, when drinking these mineral waters, would be overdosing on salt!)

I live close to the Central Highland springs area, but I choose not to drink it's mineral waters. My own preference for drinking water is a natural spring that comes filtered through the basalt of an old volcano there. There are no CO_2 bubbles! Unlike typical underground sources of bottled water, this is a true spring that emerges naturally to the surface and flows for a while through a meadow of English buttercups before it reaches the collection point. Thus it becomes oxygenated and has the chance to outgas any nasty gases, if there are any.

Hot springs in South Australia

In north east South Australia there is an abundance of geothermal resources, found as hot water and hot dry rocks deep in the Earth that are highly suitable for electricity production. Ironically, such an eco-friendly energy source may soon be used to power the nearby Olympic Dam/Roxby Downs Uranium mine.

In only one spot in the state are geothermal waters venting up to the surface from deep below. In the remote northern Flinders Ranges region, a highly faulted area, is found South Australia's only hot springs – at Paralana, 27 kilometres north of Arkaroola. These are very hot springs indeed, with near-boiling water (some 62 degrees Celsius) flowing naturally from the ground. The area is being currently prospected for potential hydrothermal power generation due to hot, highly radioactive granite below. The waters are also radioactive, with so much radon detected there that it is only safe to visit if there's a wind to blow the gas away!

ABC TV's 'Catalyst' program reported from Paralana in 2002 that:
'The amazing hot spring belches out carbon dioxide, helium and radon. This toxic soup would kill most organisms but, amazingly, there is life in the radioactive water ... A mat of green velvet slime harbours "extremophiles" that have never been found naturally anywhere else in the world.'

The radioactive slime was sent to the Astrobiology Laboratories at Macquarie University for analysis and some creatures subsequently identified in it are so different from other single celled organisms that scientists have created a new biological kingdom called Archaea (being archaic) for them – a name suggesting that they may well have been among the very first organisms to appear on Earth several billion years ago.

Roberto Anitori, a microbiologist at the Australian Centre for Astrobiology in Sydney, has found Cyanobacteria – a form of bacteria that can photosynthesise – amongst nine species of bacteria there, three of which are completely new to science. Anitori explained that:

'We are the first research group in the world to describe a bacterial and archaeal community living in a radon-rich thermal spring. The presence of radon makes Paralana unique and argues for its suitability as an analogue for ionising radiation environments, which may have been common on the early Earth and Mars. It's going to tell us a lot about how life might have evolved on the early Earth.'

The local Adynamantha Aboriginals used the pools 'for domestic purposes', Jennifer Isaacs writes, and they bathed in them to cure minor aches and pains. The springs were highly sacred and were said to be where a warrior fought for the love of a maiden. When he was victorious, he plunged his wooden fire stick into the spring, making it forever hot.

Meanwhile, over at South Australia's Mulka Station a hot artesian bore has been producing a maximum 20 kilowats of power for use on the cattle station since 1987, Victoria's Department of Sustainability informs us.

Mineral springs in Queensland

With Australia's oldest volcanic area found in north Queensland, you would expect to find some hot mineral springs there. However, only a few spring to mind (excuse the pun!).

Innot Hot Springs

In inland north Queensland, Mt Garnet and the Ravenshoe areas are popular places for visitors to try their luck at gem fossicking. Between the two areas one finds Innot Hot Springs, at Innot on the Kennedy Highway. Innot's famous hot springs are considered to have therapeutic properties and in the past the water was even exported to Europe. When it bubbles up out of the ground it can reach 75 degrees Celsius.

Nowadays tourists can enjoy the waters in one of several public pools, or alternatively dig a little hole in the sandy bottom of the creek that runs through the town. There, the mix of hot spring and cool creek waters should make for an enjoyable dip in more natural surroundings.

Yowah and Helidon Spa

Over in Queensland's central south, the small mining community at Yowah, west of Cunnanulla, are fortunate to have a hot spring bore bath. Like relatively nearby Lightning Ridge, the water is sourced from the Great Artesian Basin and is a haven for opal miners and tourists to enjoy.

In the south-east is located the spa water area of Helidon, in the Great Dividing Range near Toowoomba. This has historically been used for commercial supplies of mineral water, too. Many bores are still in use, generally for cattle feeding. A Mr Loraid set up a bottling plant there, calling it the Helidon Water Company, says the web magazine '*Old Fix*', 2006, adding that:

'According to tradition, Aboriginals knew of the supposed benefits of the spring water at Helidon. They believed that the water from the spring made the sick strong and the strong stronger'.

Researchers at the Centre for Medical and Health Physics, of Queensland's University of Technology, have been checking the radioactivity of the Helidon waters:

'The hydrogeology of the area is complex. The spa water aquifer system is located at the junction of two geological units differing in age and formation. Results of measurements of the spa water indicates the water originates from the aquifers of the Great Artesian Basin.

'[Radioactivity of] the spa waters, determined by a special liquid scintillation counting technique, lies in the range 0.5–1.3 Becquerels per litre, exceeding the recommended drinking water guidelines for livestock. At one site the overland flow and ponding of water has improved the water quality through the sorption of radioactivity onto the sediment. Further research is in progress on the potential benefits of this natural filtration method.'

Wetlands and radioactivity

'Sorption' means the binding of one substance by another. Could nature know how to deal with radioactivity? One place to confirm this is the region around Chernobyl, where radioactivity spewed over the surrounding wetland region and far beyond, on that fateful day in April 1986.

A major wildlife habitat was at risk as a result, the *New Scientist* reported on 29th December, 1990:
'*The Pripyat or Pinsk Marsh covers some 15,000 square kilometres, including the reactor site and its cooling reservoir, and it extends more than 300 kilometres to the northwest, mainly along the fringes of the Pripyat River and almost to the Polish border. The marshes are on the main flight path of waterfowl that winter in southern Europe, northwest Africa, and the Middle East.*'

University of Georgia scientists have had a major ongoing research effort in the area and have, with others, established the first baseline of what has occurred both to humans and wildlife because of the accident, Sean Henahan writes.

In 1996 researchers, including Dr Cham Dallas, reported on studies of 400 carp from 12 ponds in contaminated areas of the Ukraine some 20-30 kilometres from the site of the Chernobyl explosion. They choose fish because they knew that nearly all radionuclides – products of a nuclear reaction – go into the bottom sediments when they fall on rivers or lakes, where they are taken up by fish. They looked for only one radioactive contaminant – $caesium_{137}$, which has a half life of 30 years.

The team found numerous genetic changes in the DNA of many of the carp; however, the team found no morphologic alterations. 'The most impressive thing ... is that we didn't see more changes than we did.' The team has also noted that 'most animals studied so far in the vicinity of the Chernobyl accident show no outward changes over many generations in the decade since the accident ... It just shows the amazing resilience of nature'.

Coal seam water

In another corner of the Surat Artesian Basin, in southern Queensland, lies the drought-ravaged agricultural region of the Darling Downs. Here there's a boom happening with the discovery of enormous deposits of coal, gas and water in the Basin. The region has quickly become the third largest coal and gas mining area in the country, coal being the nation's is being drawn off from the coal seams and now supplies 50 percent of Queensland's natural gas needs.

To get to the gas in the coal, rigs have to drill down to between 400 and 700 metres. Now Queensland Gas has agreed to sell the Chinchilla Council 3000 megalitres each year of gas water as well. It's an Australian first and possibly a world first, to use this water.

Southern Queensland's drying climate has forced Chinchilla to have Level Four Water Restrictions four times in the last 12 years. And even though the gas water must be treated because it's salty (a desalination plant will be built), at 4000 parts per million it's okay to add to dams for stock water. (Stock can handle up to 8000 ppm).

Queensland Gas is running irrigation cropping trials with its gas water, adding gypsum and sulphur to counter the salt and initial results are very promising.

Hot springs in the Northern Territory

Tjuwaliyn / Douglas Hot Springs Nature Park
The Douglas Hot Springs Nature Park, on the Douglas River, is where thermal pools have created an oasis amidst dry woodland. It's a popular spot for people as well as attracting a wide variety of birdlife during the day and bandicoots, quolls and flying foxes at night. The park, 200 kilometres south of Darwin along the old Stuart Highway, is owned by the Wagiman people and is managed jointly with the Territory's Parks and Wildlife Service.

Where the warm waters of the springs join the Douglas River is the best place to bathe, remembering to check the temperatures before you enter! Water temperatures can exceed 60 degrees Celsius and swimming is better in the cooler river pools, some 200 metres upstream and downstream, Parks publicity suggests.

Katherine Hot Springs
On the banks of the Katherine River near Katherine, some 450 kilometres south of Darwin on the Stuart Highway, you can take a refreshing dip in Katherine Hot Springs. These are natural thermal springs and entry is

free. The pools are complemented by picnic grounds and scenic walking tracks.

Mataranka

South of Katherine is the small township of Mataranka, famous for being the setting for the book about life on Elsey Station by Aeneas Gunn, *We of the Never Never*. It is also famous for the Mataranka Hot Springs. These are a series of waterholes fed from underground thermal springs, and set amongst an oasis of luxurious tropical vegetation, of lush paperbark and palm forest. The tourist blurb sounds alluring:

'The warm, crystal clear waters of this natural pool have the power to rejuvenate bodies tired from a long day on the road.'

Yes, it is gorgeous, but you also must be very careful – tropical environments breed up lots of nasty bugs and the hot waters of Mataranka are no exception. I found this out for myself. After a friend and I enjoyed swimming at Mataranka in 1988, the friend ended up with an awful ear infection. He had dived under the water, whereas I hadn't. Putting your head under water is never recommended and hence the use of chlorine in public bathing pools. Some time later a nurse tending his rotting ear told us that a great many of the Aboriginal children in the Top End have these sorts of infections and it often sends them deaf and causes all sorts of learning difficulties at school. She sometimes has to syringe live maggots out of the poor children's congested ears.

Poppy's Pools

These thermal pools are situated 70 kilometres from Cape Crawford on Bauhinia Downs Station. Cape Crawford is 100 kilometres south west of Borroloola- a major base from which tourists explore the Gulf of Carpentaria and Arnhem Land regions. The Savannah Way, an adventure tourism drive along the Carpentaria Highway, is a great option for exploring this unique region, Ernie Dingo enthused, on a *Great Outdoors* TV program. Of Poppy's Pools he says that:

'Cape Crawford Tourism can arrange for a tour to visit to the little-known local gem, 'Poppy's Pool', which is owned by the traditional Aboriginal owners. Permission is required for non-local people to visit there.

'The hot spring water rises under pressure from a large aquifer system 5-7 kilometres deep in the Earth's crust. It emerges to the surface where it cools as it moves downstream and mixes with the water from the cool spring'.

Western Australian hot spring

Zebedee Hot Spring
In Western Australia's north, in the Kimberley region, is found possibly the only hot springs in the state. El Questro Wilderness Park, on the Gibb River Road, south of Kununurra, comprises one million rugged acres, including many craggy gorges. Here visitors can enjoy the Zebedee Hot Springs and relax in the hot pools available there.

Photo: Hastings hot pool

Tasmanian hot spring

Hastings thermal springs pool
At Hastings, south of Dover, tourists often make a beeline, turning off the main road south to follow a 20 kilometre dirt road, to see the spectacularly decorated dolomite caves there. The Hastings Visitor Centre and thermal pool, opened in 2004, is where cave tour bookings, pool tickets, a cafe and detailed information are available.

Beside the Visitor Centre, and amidst lush rainforest with lovely tree ferns, you can take a dip in the geothermal swimming pool and also enjoy a short forest walk alongside Hot Springs Creek, where platypus can be seen. At one point a viewing platform suspended over the creek allows you to watch hot water rising up from the deep, resplendent with gurgling bubbles, as you soak your feet.

I was told by a park ranger that the hot water, which rises at many places in the area from deep down, has been filtered through dolomite and thus contains high levels of calcium and magnesium. My skin certainly felt good and slightly crusty afterwards, following a recent swim there.

A basalt blob on the geological map tells me that a volcano was active in the vicinity, back in the Jurassic age. The geothermal waters at Hastings are around 30 degrees Celsius, says Tasmanian geohydrologist Leaman, and to be this temperature they must have come up from about one kilometre below.

Once popular with Tasmania's Aboriginals too, the Hastings hot springs area was at the boundary of four tribal areas. These first peoples no doubt shared and made good use of the hot waters in the cool temperate climate there. A Tasmanian Aboriginal woman has also told me about a warm spring pool in another part of the state that is a sacred site for secret women's business and is still in use for ritual purposes.

New Zealand's hot springs

Last, but not least, in fact most numerous, are the hot springs of New Zealand. New Zealand even has a Hot Springs Beach, on the Coromandel Peninsula, where you can dig a hole in the sand and warm up before rolling into the surf to cool off! New Zealand is fortunate in being able to produce 75 percent of its total energy requirements from geothermal sources.

Hanmer Springs

There are too many hot springs in New Zealand to mention and I shall only speak of one, which I have myself visited. Hanmer Springs is one of the most famous of New Zealand's hot mineral springs. It has been used therapeutically by white people since 1859 and has been especially popular for treating wounded soldiers, alcoholics and the mentally disturbed.

Situated in the alpine village of Hanmer Springs, 90 minutes drive north of Christchurch, the Hanmer Springs Thermal Pools and Spa has been popular for over 125 years and was voted best visitor attraction in 2004, 2005 and 2006, says the publicity. The centre is a year round holiday destination offering a variety of outdoor activities, cafes and accommodation.

When I told people in the North Island several years ago that I would be driving around the South Island several of them suggested that I go to Hamner Springs. So one day I arrived there, but I actually approached with a little trepidation.

'What facility would you like to use?' the receptionist asked me. There are several pools of different types, including nine open-air thermal pools, three sulphur pools and four private thermal pools, as well as a sauna/steam room; with pool water temperatures ranging from 33-42 degrees Celsius to choose from.

'I'd like to have a look at what you've got here, but I'm not sure if I want to go in. The last time I went into a mineral pool (at Hepburn Springs, Central Highlands) it had just been chlorinated heavily and it was very unpleasant,' I said. They scoffed at the idea of despoiling the water.
'Ahh! Those spa people in Australia don't have their act together! We don't have to chlorinate the water here and we certainly wouldn't want to!'

Thus reassured, I went in and was well impressed. The setting is so beautiful, with outdoor open-air spa pools naturally landscaped, big granite boulders all around them, clear blue skies above and a mountain backdrop. Soaking in the lovely hot water, staring up at the lush surroundings – it was heaven! Except for the cloud of midges buzzing at my face. No worries, as there were screened indoor spas to escape to and the traditional Japanese Spa was fabulous too!

Undersea hot springs beyond New Zealand

The discovery of hot springs on the sea floor almost 2 kilometres below the surface, southeast of Port Vila, the capital of Vanuatu, by a CSIRO-led team of international scientists, was reported in September 2001. An amazing array of life forms were discovered in the hot springs.

The international expedition was part of an ongoing study of the South West Pacific 'ring-of-fire', looking to discover how mineral deposits rich in copper, gold, zinc and silver are currently forming on the seafloor.

A highlight of the mission was the on-board high-precision analysis of methane in samples of deep sea water, methane often being present in some submarine hot springs. The scientists reported that:
  '*The size of the hot spring field is impressive. It's as large as the Melbourne Cricket Ground. ... The hot springs host abundant life forms living on recently erupted black volcanic lava flows and yellow-brown deposits rich in iron and possibly other metals, such as copper, zinc, gold and silver. Pink anemones, spaghetti-like tube worms, mussels and galatheid crabs are just some of the myriad inhabitants of this strange undersea world.*'

In May 2002 similar findings were reported closer to New Zealand, by New Zealand's National Institute of Water and Atmospheric Research. The media was told that:
  '*For two weeks this month, scientists investigated 13 newly mapped undersea volcanoes, dotted along 500 kilometres (310 miles) of seafloor north east of New Zealand. ... The researchers wanted to find new submarine hydrothermal vents and their associated chimneys, or "black smokers", which spew superhot mineral rich waters into the surrounding ocean.*'

The team made some startling discoveries. Marine biologists say they turned up bacteria and animals new to science living around the vents. Microbiologists recovered microbes from seawater in the 'black smoker plumes' and grew them at 70 degrees Celsius in laboratory conditions on board the New Zealand research ship.

'Extraordinary organisms have evolved around hydrothermal vents, with ecosystems built on chemosynthesis, not photosynthesis', the team said. Included in the haul are a brand new bivalve and two new species of mussel. At one undersea volcano, the scientists dredged up rocks which, when hit with a hammer on board – 'would break open and all this incredible steam would come out'. They were found to be a hot 57 degrees Celsius, said Dr Cornel de Ronde, a marine geologist with New Zealand's Institute of Geological and Nuclear Sciences.

The scientist team were able to build images of more than 50 new volcanoes, with 13 being 10 kilometres (6 miles) or more in diameter.

They found one volcano over 20 kilometres (12 miles) in diameter and 2.5 kilometres (1.5 miles) high. Long stretches of the Kermadec Arc, the boundary between the Australian and Pacific plates, are hydrothermally active, while, intriguingly- others are not, they have found. They concluded that:
'The volcanic processes to form such features must been very impressive.'

These hot water vents could well be a window to the birthplace of life on Earth (and Mars too). And they help to explain the vast quantities of sea water and slowly, but constantly, rising ocean water levels.

References

Kasting, J. F., 'Origins of Water on Earth', *Scientific American*, 2003.
Map of the Great Artesian Basin: www.environment.gov.au/water publications/environmental/rivers/pubs/gab-map.pdf
Abey, Matilda, 'Licence wait delays bores, farmers desperate for stock water', *The Weekly Times*, Victoria, 20th December 2006.
'Update on Totten Field (the Reiss Foundation's Field Laboratory)', *American Society of Dowsers Journal*, March 1986.
Webster, Richard, *Dowsing for Beginners*, 1997.
Bird, Christopher, *Divining*, Macdonald & Jane's, USA, 1979.
Water Dowers Manual, 1963–1988', American Society of Dowsers, 1990.
Wishart Edward and Maura, *'The Spa Country – a field guide to 65 mineral springs of the Central Highlands, Victoria'*, self published, 1990, (now out of print)
Archer, John, *Twenty Thirst Century – the future of water in Australia*, Pure Water Press, Australia, 2005.
Roberts, Pauline, 'Down Under Dowsing', *Dowsers Society of New South Wales Newsletter*, vol 13 no 2.
'Taking the waters in old Leeds', John Billingsley, *The Wellspring Journal*.
'Chernobyl's radioactive waterfowl', *New Scientist*, 22–29th December 1990.
Henahan, Sean, 'Chernobyl: Wildlife Follow-Up', *Access Excellence* www.accessexcellence.org/WN/SUA08/cher996.html
www.abc.net.au/science/slab/groundwater/default.htm

Willis, Paul, 'Extreme Slime' ABC TV: *Catalyst,* 3rd October 2002.
Innot Hot Springs – www.herberton.qld.gov.au/
Courtney, Pip (reporter), 'Resources boom revitalises Queensland's Darling Downs', *Landline,* ABC TV, 1st October 2006.
Olympic Dam Uranium Mine – Friends of the Earth urges people to write to Premier Mike Rann and other ministers to express concern about BHP Billiton's legal privileges and urge them to amend the *Indenture Act* to ensure the "strictest environmental standards" there. A form letter is available from www.geocities.com/olympicdam
Ponder, Winston, 'Ancient mound springs under threat,' at: www.sea-us.org.au/roxby/springsdrying.html
Rob Gourlay's website – www.eric.com.au
Bill Cox/primary water – www.primarywater.com/
www.primarywater.com/Drilling-for-Primary-Water.html
Endersbee, Prof Lance, *Monash University News,* 23rd Nov. 2005.
Endersbee, Prof Lance, *Voyage of Discovery,* self published, Australia, November 2005.
Lightning Ridge Bore Bath
www.minerals.nsw.gov.au/prodServices/minfacts/minfact_95
Paralana Hot Spring
www.petratherm.com.au/projects/paralana.htm
www.abc.net.au/catalyst/stories/s692473.htm
Isaacs, Jennifer, *Australian Dreaming,* Lansdowne Press, 1980.
Helidon Springs – www.arps.org.au/Journal/RPA14-4.php#RADIOACTIVITY%20IN%20HELIDON%20SPA%20WATERS
Radiation Protection in Australia, Journal of the Australian Radiation Protection Society, Volume 14 Number 4 (October 1996).
Helidon Spa – www.users.bigpond.com/colemanpeter/old_fizz_003.htm
G. Koulouris, J. Dharmasiri and R.A. Akber, 'Radioactivity in Helidon Spa Waters', *'Radiation Protection in Australia'* 1996.
Geothermal energy in Victoria
www.sustainability.vic.gov.au/www/html/2105-operating-geothermal-generators-in-victoria.asp

1.4 Water Divining

Dowser finds underground river

The farmers of Marrawah in north-west Tasmania have long had a theory that an underground river flowed below the ground there. Marrawah's sandy soils are useless for dam building and there are no surface creeks or rivers in the area. Many people have been searching for the underground water over the years, but without success.

However in early 2006 water diviner and Marrawah publican Peter Benson found the legendary subterranean river as he dowsed for water on the property of local shire mayor Ross Hine. Benson said that:
   'You simply cannot tell from above ground what is happening below. There's fresh water on top of Table Cape and on the Nut ... I can't find water where farmers want it all the time. I can only find it where it is.'

Using two copper-coated steel dowsing rods and a great deal of patience, what he did find surpassed all expectations. The underground river is 30 metres (100 feet) wide and runs for many kilometres, from near Dismal Swamp and into the sea. A hole has been drilled 40 metres down on Mr Hine's property to access the river below and around five million litres a day (or 1825 megalitres a year) is coming to the surface without the aid of a pump. The water is worth millions of dollars and its discovery has caused much community excitement, the local paper reported.

Source: *The Examiner*, Tasmania, 26th March 2006.

No surface water? Try water divining!

In times of drought there are often upbeat stories in the rural press about farms saved by the ancient art of divining. Divining, also known as dowsing, has been used since time immemorial to locate underground water supplies. Yet oft times in the media it is rubbished as some baseless superstition, ignorance usually being at the basis of this assessment.

Despite some people's reactions of incredulity and sometimes outright hostility towards dowsing, it has proved its worth by providing countless supplies of the precious liquid to thirsty communities, as well as the minerals that are found in this way.

Traditionally, Australians predominantly drink river water, or rainwater collected in tanks and reservoirs. But during droughts many people take the plunge and drill a bore. Often they are rewarded with copious amounts of drinkable water. But deep bores are not cheap. Fortunately, some drillers are confident enough to offer a very good business proposition to their clients. It's called 'No Water – No Pay'! They are able to offer this service because they are such adepts at water divining that they rarely strike a dry hole.

One such firm is Centrestate Drilling based in Castlemaine, central Victoria, near where I live. They even offer a free appraisal. By dowsing, of course! Old Jack divined successfully for years before retirement and his son Dan is a crack diviner as well. Between them they have a vast level of local experience in geology and hydrology. It's always better to get someone else along to dowse your own place and I was lucky to have Jack visit, as his son was on holidays.

Jack's pair of plastic rods, held in tension between his hands, his palms upturned, were lively in his hands, bobbing up and down to indicate the presence of underground flows of water. He found all the spots I had previously divined but not told him about, as well as a few more that were worth considering. (It is common to get a second or third opinion before anyone decides where exactly to sink their bore.) Jack quipped: 'of course you would find all that water, you're the one who really wants it'. Well yes, his dowsing search faculty was bound to be much more impartial than mine and not loaded with wishful thinking!

Jack also found some other Earth energies on the property that I hadn't told him about. And when he pointed his rods at the neighbour's property, they were soon bobbing up and down to indicate good potential bore sites there too. Other water diviners use an even more remote dowsing method – tracing the paths of the underground streams across a map or photo from the comfort of their home. (Perhaps the hydrologists, who don't usually dowse, are secretly 'green with envy'!) Even the beginners who I teach to map dowse often make accurate finds.

An old diviner once suggested why water diviners have been so keen and quick to develop their ability in times past. He pointed out that:
'We used to have to hand dig our wells. You don't want a dry well after all that effort!'

There have certainly been many extraordinary finds of water by diviners in all sorts of unexpected places, such as on mountain tops. Proficient dowsers have to be good also at avoiding underground waters that are often highly saline, to find the rare fresh flow, as is often the case in southern Australia.

So why aren't dowsers employed by government authorities in Australia, as they sometimes are elsewhere in the world? I imagine that if dowsing was recognised as a valid activity, it would open up a 'can of worms' in relation to recognising the true degrees of sensitivity that humans have to radiation. Levels of electro-magnetic-radiation exposure would have to be reassessed and greater precautions taken, thus inconveniencing the industries involved. Homes built under giant powerlines would have be demolished and forget about all those mobile phones! (Although perhaps many people would seem to prefer a brain tumour to giving up their phones.)

The dowser's art is certainly more accepted in some places, such as eastern Europe and Canada, where some governments have actually employed dowsers to find water. In Canada in the 1930s Evelyn Penrose, daughter of a Cornish water diviner, was the first of three government appointed dowsers hired to locate oil and water resources for the state of British Columbia. During 1931-1932, and following a prolonged drought, she also located 392 water wells for private homesteaders. Often she simply divined with her hands alone.

In one recollection from her stint in British Columbia, described in Penrose's 1958 autobiography, she was hired to help orchardists in the Okanagan region. She recalled one of her first jobs there thus:

'It was a great shock to see his orchard, covering the side of a large hill, wilting and dying ... facing disaster. We stopped and looked up the hill and he was telling me something when, suddenly, I was nearly thrown off my feet. I grabbed his arm to steady myself. "Water" I gasped. "Water! Lots and lots of water". I can never stand over underground water without being swung about, and the greater the amount of water the greater the reaction.'

Water was found only 1.8 metres (6 feet) beneath the surface. Drilling down further to 3.6 metres (12 feet) deep produced 108,000 gallons a day. The locals referred to it as the 'Wonder Well'.

In Germany and German speaking countries a higher level of awareness of the problem of 'Earth rays' has seen the employment of dowsers by government authorities for investigation into the problem of geopathic stress, which occurs when people are sleeping over 'irritation zones', such as above underground streams and geological faults. This has involved holistic team work combining physicists, geologists and other scientists as well as dowsers.

But in Australia the potential of dowsing is usually dismissed. Typically we either just accept and use it, or else it causes upset because it challenges belief systems, both of religion and science. The city population rarely come across it. Hydrologists would be challenged the most by dowsing. Hydrologists don't normally recognise the existence of underground streams, so it's no wonder they are unable to embrace this stone-age technology! They need to reassess a lot of their groundwater theories, which are not adequate to explain the presence of all underground waters, nor the presence of dryland salting.

Dowsing itself has never really been at odds with science, however. Dowsing is scientifically explicable, to some extent, and produces repeatable results – making it a useful research tool. Internationally, many scientists accept and use dowsing themselves. For instance in Russia, geologists often practise geological dowsing. Albert Einstein used to enjoy dabbling in dowsing too.

How do dowsers find water?

Dowsing is an activity in which we use a wide part of the spectrum of sensory faculties that we share with animals and that are involved with basic survival instincts. Brainwave studies of dowsers, such as by Dr Edith Jurka in the 1980s, show that we all sub-consciously sense the energy field of moving water, with our brain waves registering its presence. Dowsers have simply trained themselves to become more consciously aware of the impulses received by the body's antenna systems.

Many animals are known to be adept at finding water, the elephant in particular. Elephants have been observed wandering far from feeding grounds, to places where there are no surface indications of water, and to make a hole in the ground down to water using their trunk, thus creating a well. African elephants have been seen to plug these little water wells up after drinking from them, using wads of chewed bark, presumably to keep them clean.

Elephants appear to combine a high degree of intelligence as well as sensitivity to vibrations. They sensed the coming of killer tsunami waves of Boxing Day 2004, as many survivor's stories attest. Tame elephants on the Thailand coast tore themselves free from tethers and fled to higher ground, well in advance of the giant waves. Their keepers followed suit and all were saved from certain death. Later, and only when it was safe, the canny elephants went back to the beaches to rescue people, all at their own initiative, until everybody was safe.

Human water diviners use a variety of simple instruments to detect the presence of moving water. With them, they pick up on the energy flow paths that are associated with underground streams, rather than the more static water table. Friction created by the water's movement as it flows through geological pathways within the Earth creates an electro-magnetic field, which changes the background radiation patterns and is detectable by scientific instruments. Dowsers can easily detect these flow patterns, even down to the most minute magnetic variations.

To indicate an underground stream the dowsing rods, forked sticks or wands either bob up or down, or spread apart or come together, while pendulum movements will change to a rotation or oscillation, depending on the technique used.

Dowsers ideally need to gain a good concept of local geological conditions in the country where they are divining for water supplies. Water behaves differently in various types of rock and soil strata, and in different seasons and moon phases. Dowsers need plenty of real-life practice to develop expertise, ideally with an apprenticeship or at places were bores pre-exist and the water's depth is already known.

American master dowser Jack Livingston gained enough confidence in water divining to be able to practise his dowsing from the passenger seat of a car. His y-shaped rods dip down when water flowing under the vehicle is indicated. He also points the rods at distant topographic features and gets remote indications while driving.

Other dowsers have dispensed with instruments altogether and use their hands alone, with their palms scanning downwards or sideways, as direct sensory receivers. Others feel the energies of water with their eyes, or with other parts of their body that register nervous reactions to water.

What water diviners look for

Diviners sometimes find what are variably called 'water domes', 'sources' or 'blind springs'. This is where a shaft of water is gushing upwards vertically under the ground. It does not come to the surface, but horizontal rock fractures carry the water outwards in the form of underground streams. Historically the locations of 'blind springs' have often been associated with places of great power and sanctity across Europe.

For dowsers, this flow pattern manifests as several underground streams radiating out from a central point, as determined by dowsing for the directions of flow. 'It will usually have a feeder stream down below at a greater depth, says Irish master dowser Billy Gawn.

Dowsers have found that bees will always prefer to make their hives above a water dome and that migratory birds have been observed to nest near to them, as have sea turtles on beaches, Maria Perry noted in *The American Dowser*, of August, 1981.

In a talk given to the New South Wales Dowser's Society on 15th November 1987, professional water dowser Paul Davis (deceased) told of finding water domes with a dowsed diameter of some 60 metres (200 feet) and of finding eight circular 'magnetic lines' around them, as well as certain important Aboriginal sites that are located over water domes in the Sydney region. The report also said that:
'Paul himself can stand at the edge of a dome and feel the energy located within its diameter and the lack of it outside. It is noticeable that the energy from a dome has a harmful effect on trees within its influence',

Sometimes an underground stream can descend downwards vertically before flowing on horizontally. This is called a downshaft or downer.

Where two or more horizontal streams are crossing each other, a downward vortex is dowsable, and this is sometimes called a downshoot. Between the vertical and horizontal water flows, plus aquifers and the water table, it can be a complex water picture below the ground.

An ideal spot for a bore will be an underground stream with the most and best quality water available. Even better is a spot where two or more underground streams are crossing. If several streams are found to emanate from a 'dome', as shown by the direction of flow, often these streams will be the best tapping place for a bore. Jack Livingston and other master water dowsers advise against drilling directly into a water dome, for fear of damaging it.

Of course the location will also have to be accessible for boring equipment and water quality and quantity must be suitable for requirements. All the requested factors are held in the diviner's focus while the search goes on. Sometimes the process is initiated by first map dowsing from a remote location before going on site. This saves time and energy, as dowsing can be quite stressful and tiring for the dowser.

Dowsing water depth

The correct depth required for the bore will also be important to consider when divining. If multiple streams are crossing at varying depths, one needs to find the depth of the best stream.

Shallower bores are always cheaper, but if tapping into the water table one is more likely to find pollution and over-pumping can become a problem. The deep and ideally 'primary' water sources are the most sustainable ones to look for and are more prevalent in igneous rock areas.

To divine the depth that has to be drilled down to, there are many methods. These range from mentally counting the depth until a 'yes!' reaction is obtained; to counting by stomping one's foot, with each stomp equalling a unit of measurement; or by counting the number of times a bobber wand goes up and down. Divining the quantity of water, as litres or gallons per minute, can be approached in a similar manner.

Depthing is also done using the 'Bishop's Rule'. One divines for two parallel energy lines that run along each side of the water's primary energy line. The distance between the primary and secondary reaction lines is equivalent to the depth of the water in the ground.

Dowsing water quality

Dowsers have many ways to check for the quality of the water. Penrose speaks of how she could sometimes actually smell and taste the water being dowsed. The first time this ability became apparent to her was at a divining job at a church. Although she was in an exhausted state (which is not good to be dowsing in), she found that there was lots of water present. But she also found herself suddenly in an uncharacteristic foul mood, with a smell like rotten eggs in her nose and a nasty taste in her mouth. She stormed off and cried uncontrollably, which was also very uncharacteristic.

Meanwhile the priest told her colleague that years before a well had been dug very close to the spot she had found. The water was very high in sulphur and it was filled in. It could be that she was receiving a homeopathic dose of sulphur, as it is used as a cure for irritability, amongst many other things. ('Provings' of such remedies produce their symptomatic effects in healthy people, whereas the sick will get a cure.)

Water quality can be very variable and one needs to check for minerals, chemicals, bacterial pollutants and the like. In Australia many underground streams are saline. Dowsers in Australia have long wrestled with the salt problem. How to differentiate between fresh and saline waters?

Dowser Pauline Roberts was told of an old method that involved placing a copper coin, or equivalent, on the tongue while dowsing. Queensland dowser Veronica Hansen told her that:
'I place a piece of copper pipe on my tongue. If there is salt in the water, then the wires will go dead.'

South Australian water diviner Stanford Woolford's method was to put a small container of salt in a shirt pocket as a 'witness'. His rod would then reacted to salt water streams underground. Stan also used different colour samples as witnesses of different types of salts present while he was water divining, it was reported by the South Australian Dowsers Club in June 1986.

Some pitfalls of water divining

While some master dowsers, like America's Bill Cox, can boast of a 90+ percent success rate of finding water in all types of terrain, not all water diviners are always so successful. For instance, lunar cycles influence the energy of underground water flows. One American dowser, following 20 years of observations, reports misleadingly stronger reactions to water veins he gets on days just prior to a new moon.

The presence of a lot of heavy clay is widely felt to be a hindrance to water dowsing, disrupting energetic influences and also not releasing water to a bore. American master dowser Jack Livingston wrote about what he refers to as 'ghost veins' after he advised on two wells which failed to deliver the dowsed quantity of water. He explained that:
'Apparently, water has a hard time travelling through clay and gives the illusion of more volume than actually exists. Possibly the veins we picked, before turning to clay, did carry more volume.'

Water dowsers report another problem resulting from the presence of clay. Livingston continues:
'I know that boring with a rotary bucket in clay type material will seal the sides of the bore. And water – always taking the line of least resistance – will travel around the bore. A good many potential wells are lost this way. The old wells, dug by hand or with old augur drills, were dug more reliably, more accurately, though of course much slower. Now these big new jobs [bigger and more powerful percussion drills] can actually shatter the vein we're aiming at, divert it or drive it deeper.'

Another water diviner who became a well driller, Emanuel D Hough of Ohio, USA, has observed other such well failures with gung-ho operators spoiling what could have been a good well by not taking the time to work carefully or properly. He concludes that:
'irresponsible well drillers can, either intentionally or unintentionally, give water dowsing a bad name'.

Dam leaks?

Water divining provides other benefits in relation to storing and conveying water. When water pipes, dams or reservoirs are found to be leaking, it can be a difficult or costly exercise to seal them. Many dowsers have a good track record in locating the exact spots where the leaks are occurring, so that solving the problem is then much quicker and less costly.

Sandy Griffin, an Irish-Australian dowser, has had good success with fixing several leaking dams and is shown in the photo with his rods dowsing around his bush dam, which used to have this problem until he dowsed the leak's location and mended it.

Well gone dry?

Photo: Sandy Griffin

Underground streams of water don't usually wander, especially when their paths are determined by geological structures such as the fault lines and fissures in rock strata. But sometimes Earth tremors or earthworks with dynamite or jack hammering can result in water being diverted away from bores and wells. Or the actual drilling of the bore may alter the stream.

Fortunately, some dowsers have tackled the problem and specialise in a technique that aims to redirect underground streams that have moved and are now not flowing into wells. These dowsers also find employment in diverting veins from seeping into basements.

The technique involves inserting a large metal rod (or crowbar) into the ground at the dowsed location near the dry well (or basement). The bar is then struck hard several times in the direction dowsed for redirecting the water flow to. (I imagine that geological conditions would rule the possibility of success here.) Jack Livingston believes that these diversions are effected by the sonic waves created.

American dowser Les Tenold advises to first check if the stream targeted is shallow enough to reach the well and whether there is any rock in the way that may be too impermeable for water to get through. Tenold dowses the exact spot for diversion and drives in a heavy iron stake about one metre (3-4 feet) long into the ground there, until only the top 15 centimetres (6 inches) is sticking out. He strikes the bar with a hammer on the side opposite to the direction he wants the water to flow. Firstly, he dowses for the number of blows required.

Diversions will not usually be effected instantly. Shallow hand dug wells may take up to 72 hours before the diversion works, while a well drilled into rock can take the water from the diverted stream almost immediately, unless rock is in the way, Tenold said in *The Water Dowsers Manual*.

Other dowsers have refined the art of water vein diversion so highly that they are actually able to redirect a stream from a remote location on a piece of paper, from the comfort of their own armchair!

Village Water – the dowser's charity

The British Society of Dowsers, the oldest such organisation in the world, has its own registered charity – 'Village Water'. This is run by a group of water diviners who raise money in Britain for Zambian communities, where they go to divine for water and sink bores where needed. Gaining security of water supply can be a transforming event for the fortunate villagers. The charity will be very grateful to accept donations!

In the last two years alone, the Village Water team, supported by local volunteers, has provided water to 13,500 people in the western districts of Zambia, in the form of 79 new or rehabilitated water sources, they reported in September 2006.
   Contact: Village Water, Knoll Barn, Shifnal, Shropshire TF11 8PX, UK. Website – www.villagewater.org

Dowsing Society Contacts

Australia
* New South Wales Dowsers Society
Secretary – 21 Clontarf St Seaforth, 2092, NSW.
Website – www.divsrat.com.au/dowsing
* Dowsing Society of Victoria
PO Box 2635 Mount Waverley 3149, Vic.
Website – www.dsv.org.au
*South Australian Dowser's Club
PO Box 2427 Kent Town, SA, 5071.

New Zealand
* New Zealand Society of Dowsing and Radionics
Secretary – PO Box 41-095 St Lukes, Auckland.
Website – www.dowsingnewzealand.org

Britain
* British Society of Dowsers
The Director – BSD Office 2 St Ann's Rd. Malvern, Worcs WR14 4RG.
Website – www.britishdowsers.org (other societies also listed there)

Ireland
* Irish Society of Diviners – website – www.irishdiviners.com

North America
* American Society of Dowsers, PO Box 24 Danville, Vermont 05828, USA. www.dowsers.org
* Canada – www.dowsers.ca and www.questers.ca

References
Penrose, Evelyn *'Adventure Unlimited: A Water Diviner Travels the World'*, Neville Spearman, London, 1958.
'The Water Dowsers Manual, 1963-1988', The American Society of Dowsers, 1990.
Hitching, Francis, *'Earth Magic'*, London, 1976.
Gawn, Billy and the British Society of Dowsers Earth Energies Group, *'An Encyclopaedia of Terms Suitable for those Studying Earth Energies through Dowsing'*, August 2000.
Roberts, Pauline and Hansen Veronica, *'Copper pennies, blue dogs and lava tubes'* 'Dowsing Today', UK, September 2003.

Part Two:

Waters of power and mystery

2.1 Water's special qualities

What is water?

Water is the supreme element and the ultimate metaphor for life. In many ways behaving like a living creature, this vital liquid was a prototype and key to the development of life on Earth.

A law unto itself among liquids, many of water's characteristics are unique. It's a shape shifter that changes its form from liquid to solid to gas vapour. The only substance with three completely different forms, it can be found in the highest part of the stratosphere as well as the far depths of the Earth.

Water is the universal solvent. It can dissolve just about anything to some degree. Minerals occurring in water can give it qualities that range from therapeutic to toxic.

The basic water molecule is made of two hydrogen atoms and one oxygen atom, and the angle between the two hydrogen atoms is 105 degrees. Strong hydrogen bonds help water to retain its liquid form, despite it's low molecular weight.

Water as a crystal

Waters hydrogen and oxygen bonds are constantly reassembling themselves into geometric forms so complex that 36 different bonding combinations have been recognised. Pure water has an ordered molecular structure that creates beautiful crystal patterns, like living mandalas, as can be seen in snowflakes. Polluted water does not display the same high level of molecular structuring.

The network of hydrogen bonds can stretch and distort, making it easy for substances to dissolve, and gases to be contained, in it.

When a dissolved substance (whether it is a pollutant, or a homoeopathic remedy) is physically removed, the energetic effects may be retained as a 'memory' in the water, in the form of continued distortion of its molecular structure.

Wolfram Schwenk of Weleda Laboratories in Germany, and others, were able to demonstrate water's geometric forms using chromatography. This has apparently become a valuable quality testing method for organic food in Europe.

Japanese researcher Masaru Emoto set out to find a new technique to demonstrate the crystalline structure of water. Eventually he was able to photograph single drops of water, at 200 times magnification and at a temperature of minus 5 degrees Celsius. Healthy, living water produced photos of beautiful, clear hexagonal snowflake-like crystal patterns. 'Dead' or 'sick' polluted water is unable to produce ordered patterns and Emoto's photos of it can be 'ugly' or disturbing to look at.

Water loves curves

A liquid water drop in free fall adopts a preferred shape of a perfect sphere; while water loves to flow in circles and waves and to swirl around in spiralling vortical and curving motion. Pushing water through pipes, concrete channels and hydro-electric schemes, and forcing it to flow along straight channels, drains away it's life force and spirit (as do distillation, pressure and heat).

Water wizard Victor Schauberger discovered that rivers that had been put into straight concrete channels flowed faster and became flood prone. He was able to reduce the damaging effects by helping the water to flow more slowly and naturally, with the aid of baffles in the river bed that jutted out and deflected some of the flow in different directions, and created turbulence.

Water and energy

Natural, healthy water embodies life and carries life force. Its energetic characteristics are dependent on its movement and temperature. Fast moving and hot waters are yang in nature and are a stimulating force. When waters are calm, cool and languid, this yin state is soothing and contractive.

Water responds to energetic influences readily. Professor Callahan showed us how water is diamagnetic, that is, weakly repelled by magnetic force, such as the magnetism of the sun and moon. Experiments back in the 1950s in Italy and Belgium showed that chemical reactions occurring in water were related to sunspot activity and changes to the Earth's magnetic field. Schwenk's water drop pictures revealed the high degree of water's sensitivity and showed that it also responds to the moon and to the more subtle influences of planetary movements. Dr. Schwenk is the author of *Sensitive Chaos*, which details the great sensitivity of water to magnetic, electric, gravitational and vibratory influences.

According to EMR Labs, we now know that water is:
'... affected by light, sound and pressure. All these sensitivities are fleeting and the only time water is really sensitive to external forces is when it is flowing. The moment the flow ceases, the energy of the moment is retained in the water until it is agitated or moved again, at which point a whole new set of energies are captured.'

Water actually consists of layers that slide along each other when it is shaken or otherwise turbulent. When this happens the surfaces of the layers are most susceptible to absorbing energies.

Scientists have also discovered that rhythms of the moon are able to pulsate in all water – from the ocean tides down to single water droplets. Our brains, which comprise some 90 percent water, are thus also highly subject to the influence of the sun, moon and planets. Full moon is a stimulating time for all life and this is when we talk most on the telephone, as an Australian Telstra study has found.

Folklore often emphasises our lunar links. There are European traditions that dictate digging wells only during certain phases of the moon, sun and planets. Well water is said to rise higher at certain times; at other times it's deeper to reach.

At full moon, rivers tend to spread out and it is more difficult to float logs in them, as they may be washed up on banks, Schauberger discovered. At new moon, rivers are more contracted and logs are drawn into the middle and are more controllable. Stream currents are strongest at the earliest and coolest hours of the morning and also at full moon.

Sun and water makes natural energy

In nature sunlight splits water into hydrogen and oxygen. Now a means of mimicking this is being developed by a research team at the University of New South Wales led by Prof Janusz Nowotny, it was reported in March 2007. The team are world leaders in using titanium dioxide as a catalyst to split water in order to make hydrogen for use as a fuel. The technology works best with sea water and its only by-products are oxygen and fresh water.

The team estimate that 1.6 million of such solar devices, installed on rooftops (or an area of 40 square kilometres), would be sufficient to supply Australia's entire energy needs. The technology, which was first developed in Japan in the 1970s, relies on a light-sensitive photoreceptive titanium oxide cell to harness the sun's energy. Unfortunately the invention is expected to take another ten years to develop to a commercial level.

Hydrogen as a fuel is clean and green, and can power anything from cars to air conditioners. When burnt it gives off water. Germany and the US already have hydrogen refuelling stations for cars. And since September 2004, Perth has trialled three Daimler Chrysler hydrogen-powered buses, adapted from diesel fuel, in one of the first major trials of hydrogen fuel cell buses in the world. Daimler Chrysler hopes to be commercially producing the hydrogen buses in Germany by 2010.

Why this simple, clean and green technology isn't being developed more quickly is a mystery. We needed this yesterday!

Temperature differences

Water's abilities change with temperature. Unlike other liquids, water bodies are warmest at the bottom and coldest on top. This means that lakes and rivers don't completely freeze over and life can go on under the frozen surface.

The biological zero of water is +4 degrees Celsius and it achieves its maximum density and heaviest weight at that temperature. It also has the highest energy levels and biological quality, and the least potential for bacterial contamination at that temperature. Colder than that and water starts to expand and is lighter than its liquid form. This is how icebergs are able to float in the sea.

Water moves fastest when it is cool. The most highly energetic water is found in the coolest places, or at night and in the early morning hours. The guardians of sacred springs and wells knew to keep their sources shaded from heat and protected by vegetation, or, in later times, by structures of stone. When water is cool, its memory is heightened. And one of the healthiest types of water to drink is said to be glacial meltwater.

At human body temperature, 37.5 degrees Celsius, water has a maximum of structural possibilities and ability to acquire information, say the EMR Labs. Hotter temperatures than this and water's capacity for transfer and memory is reduced. Water's boiling point is one of the highest of all liquids and varies according to altitude.

Distilled water suffers a total erasure of information. However water condensation, formed as water cools after the distillation process, is able to absorb new information again immediately.

Energies of thunderstorms

Electrical storms happen when clouds rise high and winds blow through ice crystals, creating friction that releases massive amounts of energy from the water. Apart from the obvious manifestations of thunder and lightning, unseen energies rain down upon us during electrical storms. Gamma rays and X-rays shower down from the clouds. Above the clouds gigantic 'sprites', some many several kilometres across, discharge electrical energy in massive vertical plumes that have occasionally been photographed.

Also generated by lightning storms are low-level, but fairly coherent, waves ranging from 8 to 40 Hertz (cycles per second). These are known as Schumann waves or resonance, and they are claimed to be important to life and health. Professor Callahan likens them to our brainwaves, which are of a similar frequency, and even describes them as 'atmospheric brainwaves'.

Mega-cities today are magnets for electrical storms, because the hot air updrafts rising from them create ideal conditions for storms to occur, and with greater frequency and ferocity.

Lightning and life

When lightning bolts discharge, they ionise the air and produce nitrogen oxide. It's thought that this process may generate more than 50 percent of usable nitrogen in the atmosphere and soil. This explains why electrical storms have amazing effects on plant growth, which cannot be replicated by mere irrigation. I myself have seen plants apparently come back from the dead after a thunder storm.

Lightning storms were no doubt a cradle, or foundry, of life creation. Scientific experiments have shown that life-building amino acids can actually be created by the interaction of lightning and atmospheric gases.

Stanley L. Miller and Harold C. Urey, at the University of Chicago in 1953, took methane (CH_4), ammonia (NH_3), hydrogen (H_2), and water (H_2O) and ran a continuous electric current through the system, to simulate lightning storms. After just one week Miller found that as much as 10–15 percent of the carbon was now in the form of organic compounds, with 2 percent of the carbon forming some of the amino acids found in proteins.

In Russia in February 2007 a lightning strike on a haystack produced a bitumen-like substance – a *phytofulgurite*, the likes of which would normally require millions of years to form. It is the first time that this substance, a gooey black lump of diverse amino acids, has been scientifically described. (Fulgarites are the blasted tubes of melted sand that has been lightning-struck.)

The scientists of the Russian Academy of Sciences think that it was the stormy combination of electrical energy plus gamma and X-rays that created the thing and that also, perhaps, new life forms may be blasted into life this way.

Some icy comets that come to Earth are dark red or black as asphalt, due to large amounts of complex carbon compounds in them. Early Earth copped a heavy bombardment by comets, which provided a large supply of complex organic molecules, along with water and other essential goodies. (Amino acids have also blown in on meteorites, and 90 amino acids were identified in just one meteorite that fell in Murchison, Victoria, in 1969.)

Storms and ionisation

Globally, thunderstorms act as a worldwide circuit for recharging Earth's batteries, maintaining the atmosphere's positive electric charge. Some researchers at NASA think that the Earth's negative charge is also maintained by electric fields in thunderstorm clouds.

During the build-up to a thunderstorm, positive ions predominate in the atmosphere. Tension is palpable and some ion-sensitive people become anxious. Old wounds and arthritic joints may start to ache.

When lightning finally strikes, the positive ions are neutralised and negative ions start to predominate. When rain is falling, the clear, sweet-smelling atmosphere is blessed by an abundance of the negative ions that are so very beneficial to our wellbeing. Even if rain is slight, there's a wonderful sense of relief after rain or a dry thunder storm.

But not all rain has this effect. Monsoonal rains, whose raindrops are large in an oppressively hot humid atmosphere, actually produce positive ions which can be depressing for people, especially when rain lingers and mould starts growing inside your home!

Positive ions and static electricity are the scourge of arid or desert locations, where winds become highly positively ionised and are known locally as 'evil' or 'witches' winds. Some famous 'evil' winds are the khamsim, the sirocco and the foen. Animals and plants all sicken under their influences, and bacteria and viruses can proliferate. Not surprisingly, where these winds occur, fountains are popular central features in walled gardens, where the splashing waters can help provide beneficially high negative ion environments.

Unnatural indoor environments and high-tech living also expose us to many positive ions, and ideally we should stay grounded to Earth for the sake of our vitality and health. According to a study by Clint Ober and replicated by two other separate researchers, that is mentioned in a 2004 Powerwatch publication:

'a previously unknown natural bio-electrical shield of negative ions exists in and around the body when humans maintain physical contact with the Earth. This shield plays a vital role in protecting the bio-electrical activities of the body from extraneous electrical interference...[and] naturally neutralises positive ions.'

To help maintain a high negative ion state, walking beside raging rivers and crashing waves on the seashore, visiting waterfalls and breathing crisp mountain air are nature's ideal prescription. The perfect ratio for a human environment is 60 percent neg-ions to 40 percent pos-ions.

Water as a geopathic irritant

Then there are the highly ionised atmospheric flows found over underground water courses. The energies of these flow paths are presumably related to the dissolution of substances that have come into contact with the water, plus the energy of the moving water itself, which have altered natural background radiations. (These energy lines are sought out by diviners in their search for water supplies.)

These energy flows are high in positive ions and electro-magnetic radiation, and, as they also rise vertically, are unhealthy influences to be exposed to. Referred to in the past as 'irritation zones' or 'black streams', in more modern, scientific terminology they are called 'geopathic' and 'geopathogenic' zones. German tradition speaks of 'cancer beds' and 'cancer streets', because in Germany cancer is known to be a consequence of this sort of exposure. (The subject has been dealt with more thoroughly in my book *Divining Earth Spirit*.)

The energy flows above underground streams were described by the ancient Chinese. Chen Ssu Hsiao, for instance, wrote in the 14th century that under the ground there are alternate layers of earth, rock and flowing spring waters. The water currents flow along *ching*, channels that carry the Earth's vapours, its *ch'i*. The association of water flow paths and divining is seen in the name of the ancient divinatory classic tome, the *I Ching*, Christopher Bird points out, in his book *'Divining'*. And *feng shui*, the science of subtle environmental influences, means 'wind-water', because these elements convey energy to (and away from) the environment.

These days scientific investigation has access to instruments that can test for and measure the radiation frequencies occurring above underground water courses. The ionised fields and microwave emissions above water lines can no longer be dismissed as the delusions of dowsers!

German dowsers have also reported the presence of 1420 megaHertz radiation above underground water (German magazine '*RSG*' no. 175, January 1986). Apparently this frequency is also found reserved for radio astronomy, as it is the reaction frequency of hydrogen atoms reacting with each other. Serious research into geopathic problems seems to be done mainly in the German-speaking countries, and it enjoys government funding too.

Water's memory

The ancient Phoenicians called water *Mem*. Mem is also the thirteenth letter in the Hebrew alphabet and letter for *mayim* – water. *Mem* is the root word for 'memory', thus ancient beliefs preserve water's capacity for pattern carrying and transference.

The modern concept of the memory of water was first alluded to in 1988, when the prestigious publication *Nature* published the controversial work of Parisian biologist Jacques Benveniste and his team concerning homoeopathy. The team had found that a substance in solution, when highly diluted until no actual molecules of the original substance remains, can behave as if it were still there, the water retaining the molecules memory.

Critics were scathing. This defies the laws of physics, they cried. How can the information still be transmitted in the absence of the molecule? Yet the efficacy of two hundred of years of homoeopathic treatments are self-evident. (The English royal family, for instance, swear by this modality and are patrons of homoeopathic hospitals.)

Twenty years ago Australian scientist Dr Sergei Barsamian proved that a virus can spread through the Earth's energy field and replicate itself at a distance (*University of Sydney News*, 5th August 1986). This helps to explain how bird flu virus may be getting around, even in the European mid-winter, despite the lack of foreign bird contact.

Elsewhere, experiments by Professor Ludwig showed that homeopathic remedies could be transferred to water from sealed containers floating on the water. Radionic practitioners regularly transfer the energy of remedies into water by remote means. However it appears that the product, while replicating a substance energetically, doesn't have such a long shelf life as traditionally made homoeopathic solutions.

Water and emotions

Maintaining an emotional equilibrium is fundamental to our physical, as well as psychological, wellbeing and quality of life. Within our watery brains the strongest memories that persist over time turn out to be the highly emotionally charged ones. So emotion is a constant influence in our lives.

Water's ability to carry an emotional charge is seen in experiments conducted in the 1960s by the Canadian Bernard Grad. Water that had been held by a happy psychic healer had a good effect on the growth of seeds. Water that had been held by a seriously depressed person had a negative effect on plant growth, Grad found. Emotional healing effects are noted for some people who merely spend time in the presence of healing waters, without making any physical contact with them, or drinking them.

Emoto's water crystal photos are able to show us emotional impacts on water's structuring. Tragic events, such as an earthquake at Kobe, showed up as a shock felt by its water. But not long afterwards, Emoto was amazed to see well-formed crystals develop in Kobe's water supply. So much help and compassion had flowed into the stricken city that the shock felt by the water had been cured!

Surface tension

Layers of healthy water flows incorporate braiding and clusters, with water surfaces exhibiting a 'skin', thanks to the quality of surface tension. This skin is only a molecule thick, yet it has incredible tensile strength. It holds water together so that it can pull itself along through our body cells or be drawn up to the top of a tree with capillary action.

Furthermore, 'some scientist speculate that the "skin" of water is the basic blueprint for the membrane of living cells, which regulate the passing of vital solutions and ions in and out of cells,' say the Keegans.

The degree of wetness of water is relative to its surface tension, which is dependent on a range of factors such as temperature. The lower the tension, the wetter the water and the more easily absorbable it is. The degree of surface tension affects the capacity for energy transmission as well.

Many would be familiar with tales of the radiant health and longevity of people living in the isolated Hunza Valley in the north of Pakistan. Much of their pizzazz depends on the vibrant nature of the water they drink, water from glacial melt, which is tinged with crushed minerals in colloidal suspension and has a low surface tension.

American researcher Pat Flanagan tried to replicate this Hunza water. In the process he discovered that placing a gemstone or crystal in water can alter its surface tension. The effect dissipates from half an hour or so after the stone's removal. An ancient Tibetan medicinal practice places precious gemstones in water; the water is then drunk immediately to cure specific ailments.

Schauberger found that he could restore and re-energise water in an egg-shaped device. To replicate Hunza water Flanagan had to devise a machine for inducing a whirlpool-like vortex. The energy of the resulting water vortex was measured at up to 20,000 volts. Flanagan found that, by stirring to create a vortex, first in one direction, then the other – you create an implosion, that has anti-gravitational side effects (Christopher Bird, *American Dowser*, April 1988).

As for water's skin being the original membrane of life, I would like to take the idea further – that water, as the prototype of living cells, was actually the original first living being on this planet.

The definition of living in my dictionary is:
    'To have life like a plant or animal'... 'be capable of vital functions'... 'to continue in existence, operation, memory...'

Considering the vital energy, memory and spirit of water, it seems to me that water is indeed alive. Was there any life before water? I doubt it.

References

Sangster, Hugh, *A Coherent View of Health and Healing*, Dowsers Club of South Australia Newsletter, 1990.
'Structured Water or 'Miracle' Water . . . Another Method' EMR Labs, at: www.quantumbalancing.com/links.htm
Callahan, Phil S, *Paramagnetism – rediscovering nature's secret force of growth*, Acres USA, 1995.
Knonberger, Hans, and Lattacher, Siegbert, *On the Track of Water's Secret*, Uranus, Austria, 1995.
Smith, Deborah, 'Science Turns Sun, Surf into Green Energy', *Sydney Morning Herald*, 21st March, 2007.
Lightning:
www.en.wikipedia.org/wiki/Miller-Urey_experiment
Lightning and the Space Program, June 1990 (Revised 1995) www-pao.ksc.nasa.gov/kscpao/release/1990/72-90.htm
Murchison meteorite:
www.chem.duke.edu/~jds/cruise_chem/Exobiology/miller.html
Wallach, Dr Charles, *The Ion Controversy – a Scientific Approach*, Belle Lumiere, Australia, 1983.
Hartley-Hennessy, T, *Healing by Water or Drinking Sunlight and Oxygen – a new yet very ancient approach towards disease*, Essence of Health Publishing Co, South Africa, 1950.
Alexandersson, Olof, *Living Water – Viktor Schauberger and the Secrets of Natural Energy*, Turnstone Press, UK 1982.
Boulbik, Dr Jaroslav, 'Water Quality for Wellness', in *Options*, (Australia) issue 9, 2006.
Ryrie, Charlie, *The Healing Energies of Water*, Journey Editions, USA, 1999.

2.2 Water and phenomena

Mystery lights

Much folklore has evolved around the precious water sources of the Earth. Some wells and springs have been associated with mystery 'fairy lights', which nowadays might be called 'Earth light' phenomena. Some wells have been seen emitting beams of light. Geologically speaking, this is highly feasible. Quartz rock emits light when under pressure. As many springs come up in geologically faulted country, perhaps they are acting as inverse antennae, venting piezo-electric effects from quartz-rich rock below.

In Brazil, Earth lights are traditionally known as *Mae de Ouro* – 'Mother of Gold'. The first body of water that the lights cross is said to contain gold. Perhaps this is related to the idea that a 'crock of gold' lies at the end of a rainbow?

In the China Sea lights resembling 'Chinese lanterns' or 'globes of fire' floating in formations above the water are often seen by fishermen in the Shimbara Gulf and around Japan. They are associated with certain cold-weather conditions.

In Scotland, where much volcanic activity and Earth faulting makes for diverse geology, there are many reports of mysterious lights. Author Paul Devereux tells us that Earth lights are often reported at major dams and water reservoirs. These bodies of water are heavy enough to provide enough pressure to cause 'microquakes' – local Earth tremors. Seven out of eight Scottish lochs where mysterious Earth lights are occasionally seen are associated with major geological faults, the eighth being on a locally major fault line, Devereux explains; while there is also evidence of tectonic strain, most of these locations are seismically active.

Loch Ness, of water serpent, monster fame, is one such loch. It is located on the Great Glen Fault, which is 'possibly Britain's premier fault line', says Devereux. On Loch Tay, northwest of Dundee, there were reports of 'balls of fire' skimming the surface in the 1930s, while 'tongues of flames' were occasionally seen close to Loch Maree on the west coast. The greatest number of lights have been reported from Loch Leven, especially around the early 20th century, when they achieved some fame. Several seismic events occurred there between 1967 and 1984. Since then relatively few light sightings have been made, perhaps a result of geological pressures easing.

In Northern Ireland, the 'White Light of Crom' has been seen many times around both ends of Lough Erne in County Fermanagh; and at Lough Beg in Derry in 1912 a mystery light was repeatedly seen around Church Island, where there was a saint's shrine.

Swamps, marshes and bogs have always been regarded with fear and awe. The bodies of human sacrificial victims, some remarkably preserved, have been brought up from deep down in bogs, often as a result of peat-cutting activities.

Wetland areas are sometimes dangerous for the 'marsh gas' – methane – that is prevalent there. It can be deadly. Generated from underwater decomposition processes (and perhaps from deep within the Earth too) this marsh gas has been suggested as an explanation for mystery lights known as Will o' the Wisps or Jack o' Lanterns. But it isn't a very good explanation, for anyone who has tried to make this gas act in the same way.

Water and UFOs

Not surprisingly, the mystery lights associated with water bodies have been interpreted by some as UFOs. And perhaps some of them are more than just light. Many eye witnesses say that the behaviour of these lights appears to exhibit some kind of intelligence.

There are reports of otherworldly things that fly in and out of certain lakes. For example, the modern Sylvan Reservoir in the Dandenongs, near Melbourne, has been the scene of such sightings, I was told, and beneath its waters an Aboriginal sacred site is now drowned.

In New Zealand's Lake Ototoa, north of Helensville in Northland, a light is regularly seen shining up out of it at night, an informant tells me. She also reports seeing for herself a 'UFO' going down into the lake. It's a locally well-known phenomenon and curious international tourists also come to look at it.

Sightings like this have also occurred at locations in the sea too. Puerto Rico – on one corner of the so-called Bermuda Triangle – is a hot spot for Unidentified Submergible Objects, says author Timothy Good. In his year 2000 book on the subject, Good:
'reveals information that confirms that aliens have established subterranean and submarine bases on Earth.'

Water energies and dowsing

As well as recognising the association between underground water and geopathic zones, dowsers, particularly in Europe, find that the environments of ancient sacred sites, temples and churches are often pulsing with water energies.

Dowsers typically find underground water line crossings beneath the centres of ancient stone circles, as well as under the altars and steeples of churches – in old European churches, at least. (This subject is also described in my books *Divining Earth Spirit*, and *The Magic of Menhirs and Circles of Stone*.)

The water radiations at sacred sites probably have an uplifting effect on the people who use them, but are too strong for longer term exposure. The effect of sacred ceremonies at these places potentially spreads out into the surrounding environment, carried along by the water line energies.

Dowsing reveals that the surface waters of landscapes these days are generally energetically poor for various reasons. The effect of metal structures can disturb the subtle aspects of rivers, for instance. It's an old problem going back to the Iron Age; in some places this problem was intuitively recognised. In pagan Italy, for example, the Romans forbade anyone from building a bridge of iron over the River Tiber lest their river god Tiberinus be enraged.

The geodetic energy flows ('serpent lines') that normally follow waterways are deflected from their paths wherever there is a metal bridge, which is what most modern bridges are made of in Australia (and probably elsewhere). The spirit of water does not like iron or metal in general, although a little silver or copper can have good effects.

References

Marsh gas – www.wikipedia.org/wiki/Marsh
Holiday, F. W., *Serpents of the Sky, Dragons of the Earth*, Horus House Press, USA, 1973.
Good, Timothy, *Unearthly Disclosure*, Century, 2000.
Deveraux, Paul, *Earth Lights Revelation*, Blandford, UK, 1989.
Phillips, Alasdair and Jean, *Mobile Phones and Masts, the Health Risks*, Powerwatch, 2004, UK. www.powerwatch.org.uk

2.3 Tapping Water's Wisdom

Water and wisdom

Water's multiple connections with humankind have been evolving since earliest times. Water has provided poetic inspiration, quiet reflection and psychic revelation, and has been traditionally invoked for problem solving, illumination and healing. Not only does water interact with our own thoughts and feelings, it exhibits divine intelligence in it's own right.

The sacred Hebrew letter *Mem* is representative of not just water but, on a higher level, also, as Rabbi Yitzchak Ginsburg explains, it is:
   'the fountain of Divine Wisdom of the Torah /Kabbala ... the flowing stream ... (And) just as the waters of a physical fountain (spring) ascend from their unknown subterranean source ... to reveal themselves on Earth, so does the fountain of wisdom express the power of flow from the superconscious source.'

Water as a source of creative intelligence and inspiration has been expressed in animist traditions around the globe. In Greek mythology the function of the Muse as patron of the arts continues a tradition of Celtic water goddesses and lesser spirits of inspiration. The Muse tradition held that there were three or nine sister Muses. Hesiod cited the triple muses as *Melete* – Practising, *Mneme* – Remembering and *Aoide* – Singing. *Mnemosyne* – Memory Personified – was the mother of the Muses, whom she conceived in nine days of love making with Zeus. In Boeotia Mnemosyne was worshipped at sacred springs.

The *Naiades* are the Greco-Roman water spirits of sacred pools and springs, and these were invoked locally as patrons of music and poetry. If we 'muse' upon something we are reflecting, gazing, or silently meditating. And then understanding flows.

More on water and emotions

Water has come to represent or rule our emotions. People who are deemed 'wishy washy' are often highly emotional and when we are upset we put on the 'water works'. Tears are therapeutic, having salty biochemical stress relievers in every drop. There's a close resemblance to sea water in their composition, perhaps a signature of our ocean origins.

On the other hand one is warned in feng shui that a well can become a reservoir for sorrow and bitterness if it's poorly maintained or allowed to go stagnant. Author Sarah Rossbach advises to counteract this by adding vegetation, such as a pot plant, to the well cover to cheer up the water below.

Mythically speaking, the Spirit of Water is usually personified as a wise and compassionate woman (although water spirits can be either feminine or masculine). Water, the Moon and the feminine principle have long been linked together as yin expressions of the female. Music expresses both ideas and emotion and is presided over by the Muses, who are depicted as beautiful women.

The influence on us of the Moon is said to be stronger in the watery places of the world, according to Rudolph Steiner. So beach holidays are probably the most romantic!

While water impacts on our emotions, water also responds to our feelings. Dr Emoto has shown this and life usually bears the idea out. I remember when the mains water supply seemed to be reacting to wild emotions in my house. On two separate occasions and places, an argument with a loved one was seemingly echoed by the bursting of a mains water pipe in the street outside!

Watery transformation

Sacred water sources, as seats of consciousness and power in the landscape, not only came to personify centres of life but also transformation and death. They were also seen as entrances to the unconscious world, the depths of the underworld.

Springs were once imagined to be passageways leading to Earth's inner womb. Themes of death and rebirth via waters in the Underworld are common, and similar, world wide. In China, the souls of the dead are said to go underground into the 'Yellow Springs'.

The River Styx is the sacred waterway that has to be crossed on the path to heaven and rebirth in Greek mythology. Styx is both a goddess and a river flowing along the boundary of Hades, between Earth and the Underworld. The Styx circles Hades nine times and is one of several rivers that converge at the centre of Hades on a great marshland. The Sanzu, or River of Three Crossings, is the Japanese Buddhist version of the River Styx.

Some native north American tribes have a tradition of placing a bowl of water beside a dying person. This is to allow their spirit to go into the water at passing and from there to be reborn.

Attracting water

'Not only do the thirsty seek water, but the water also seeks the thirsty.' (*Jelaluddenrumi, 1237.*)

Aboriginal legends speak of the creation of landscape features, including the waterways, by the spirit heroes and totemic beings of the Dreamtime. As these beings moved through the land or named it, the features were created.

In traditions from remote Elcho Island, off the coast of Arnhem Land, in Australia's Top End, the Dreaming creation of the islands water sources was achieved with magical walking/digging sticks. Symbols of the creative act, several colourfully decorated sticks about 1.4 metres long, can be seen on display at the fascinating Berndt Museum in Perth.

These 'Ganinjeri' symbolise the sticks used in the Dreamtime by the two Djanggawal sisters, daughters of the Sun. They are traditionally made as a visual reminder of the creation times, when the sisters made the land as their journeyed. As well as the land, the sisters brought forth the flora and fauna, sacred objects and rituals too. And wherever they plunged their walking sticks into the ground, fresh springs burst forth. Circular markings on the sticks represent the water holes they created in the Dreamtime.

In the Bible, too, we have examples of water spontaneously appearing as a result of some inspired action. The best known case is the instance where Moses rod struck a rock and water gushed up. Perhaps

The Wisdom of Water

Moses was a diviner. It's interesting that this spring was named after Miriam, the Jewish prophet and sister of Moses. Her name is derived from the word for 'water' and she was also a master poet, says author Patricia Monaghan.

Many holy wells were often established, according to legend, by an initial spiritual act. Sometimes goddesses disappeared into the Earth at spots from where springs subsequently arose. Later, Christianised stories replaced the goddesses with saints who, through an act of piety or heroism, caused the springs to arise, which they then blessed and deemed sacred. Sometimes it was the heads of saints and heroes that were severed; a spring manifesting on the spot where they had fallen. The stories show an enduring connection between the human mind, spirituality and water.

The Spirit of Water is often personified as an attractive woman, irresistible to men. Sometimes gorgeous water spirits marry human men and make delightful wives, but there is always a tragic end to these legendary tales.

In modern times, people report experiences of the attraction of water. When regular rituals at modern stone circles and labyrinths have been instituted, for instance, underground streams have been found later, by dowsing, to be newly coursing beneath them.

Dowsers find that these shallow streams take up to three months to establish their new positions. This follows on from earlier British dowsing research (such as Guy Underwood's work) which established the energetic presence of underground streams and water domes beneath churches, and the crossing points of water lines beneath steeples and altars.

Sonic vibrations seem to be particularly influential for water. In Chinese legend certain musical masters were able to influence the elements and weather by using the vibrations of their voices and musical instruments. One musician was so adept and his zither so in harmony that 'beautiful winds murmured, clouds of good fortune came up, there fell sweet dew, and the springs of water welled up powerfully,' Maria Perry noted in *The American Dowser*, February 1985.

In India in 1982, in a drought stricken area near Madras, at government bequest a fortnight of rain-making ceremonies were performed at a

reservoir. These featured a famous violin player and composer, Kunnakudi Vaidyanathan. It was reported that:
> 'The passionate notes of an ancient Indian tune wafted across the warm waters of Red Hill Lake. When the eloquent violin music faded, an expressive chant floated into the air – "Varshaya, Varshaya".' (It did rain, but only 3 millimetres, one week later.)

Sonics possibly provides an explanation for some dowser's ability to repel water too. When underground water streams are causing damp in basements and geopathic stress on highways, some dowsers insert rods into the water line. These are then given a good few whacks in the required direction. Sonic waves are thought to assist with altering the water's pathway. Whatever the explanation, it does work!

'Sounds plus intention equal manifestation', it was pointed out in a film about Fountain International's community work. If you want water badly enough and the intention is clear, it is possible to influence it merely by loud verbalisation, according to a story told by an Irish dowser in *Dowsing Today*.

Christopher Strong wrote of life at the Mill House on the River Slaney in Ireland. A bedroom door kept mysteriously opening of its own accord at about 12.50 am each night. Cross with having his sleep interrupted, one night he sat up in bed and shouted 'Whoever you are, go away and leave us alone!'. They were not disturbed ever again.

So when the well suddenly dried up and there was no sign of any water flowing into it, Peter the handyman suggested they try what had been done in the bedroom. Strong wrote:
> 'So, giggling I shouted down the well "Could we have our water back please?" We laughed and shrugged our shoulders and said "Well that isn't going to work, is it?" Whereupon there was a gurgle in the well and a gush of water from the side of the shaft and, within no time, the well filled again. Peter turned pale and asked whether we had a vodka in the house. The well never dried up again.'

People at certain lakes in China's southern province of Yunnan simply yell for rain. The longer they yell, the longer it rains! Robert A Nelson explains that:
> 'This effect is possible because the air there is so saturated that sound waves can cause water molecules to condense'.

The Gift of Water
by Lee and Frits Ringma

Lee and Frits Ringma are practitioners of Agnihotra and Homa farming, in the Hunter Valley of New South Wales. Agnihotra is an ancient Vedic fire ritual which aims to purify and vivify the atmosphere, harmonising and enhancing all life. Their story is a wonderful illustration of the spiritual nature of water.

'When we purchased the land in 1994 it was in drought and the only source of water was harvested rainwater held in concrete tanks. With ongoing drought we had to resort to recycling our water. The bath water washed the clothes and the dish washing and clothes washing water watered the non-food trees and plants.

One day a water diviner turned up and after thoroughly dowsing the land he told us there were no underground streams to speak of. It was a spartan start on the land.

However on a trip to India to visit Shree Vasant Paranjpe, (author of *Light Towards Divine Path* and *Homa Therapy – our Last Chance*), we visited a sacred site where a large fresh water pool exists a few metres from the sea. Little did we know a drama was about to unfold. A Brahmin was taking his bath. He slipped and fell off the immersed slippery steps into deep water. He started to flounder and it was clear he was drowning. We were witnessing this from a balcony up in a temple complex with no immediate access to the water. Frits ran around a block of streets to try and access the water. By the time he reached the water's edge the man had been under for some minutes. Frits dived in, found a leg and pulled him ashore. The man was blue and not breathing. He appeared dead.

However Frits, saying Mantra, experienced a movement of energy and the man then came to.

Later, through a Shastri in Mumbai, we were told that due to this act we would find water on our land. As a result, on returning to Australia we had a bore well drilled, despite the prognosis of the diviner that there was no underground water.

Water was found at 33 metres (100 feet). It was laboratory tested and found to be highly saline and alkaline. So here was an opportunity to see what Homa Therapy could do. We did Agnihotra by the bore and regularly placed Agnihotra ash down into the bore well. The state Department of Water Resources was conducting regular tests on the bores in our neighbourhood and we were all amazed to see the salinity and alkalinity drop with each lab report until we had potable drinking water.

The story does not stop there however. Another diviner, highly renowned for his intuitive ability, recently turned up at Om Shree Dham to learn about Homa Therapy. He was interested in divining the land. After walking the land he came to us saying that the underground stream of water was acting very strangely. He said it travelled along a certain course then made a sharp 90 degrees turn, proceeded under the Fire hut and to the other side of the property then turned back on itself and continued back to the original course along the other side of the property. We asked him to show us where it turned back on itself. It had made a beeline for the borehole and then returned to its original course! We had not told this diviner that we had a bore and as a shed is built around it, there was no way he could have known it was there.'

Tears of the Earth

In the ancient scheme of correspondences, springs of water have been personified as the tears, and wells the eyes, of the Earth. Rain has been said to be tears of the heavenly gods as well.

Springs also represent the tears of the gods and mythic ancestors. They feature in two South Australian Aboriginal legends, written up in the 1920s by Narranderie author David Unaipon, whose face appears on the $50 note. Unaipon tells the story of the culture hero Narrundjeri and his wives as they created landscape features in the Dreamtime. In a journey along the Murray River Narrundjeri was in hot pursuit of his two law-breaking wives and wanted to punish them. At one point, stopping at Point Elliot/Raukkan, an Aboriginal community today, in his sadness over the situation Narrundjeri wept copiously beside a big rock, his tears flowing into the sea.

Unaipon continues:
'He wept so much that today some of the old folk will point out the place and say: 'This is the spot where Narroondari wept bitterly for his two wayward wives.' It resembles a soakage of fresh water by the side of the sea, and that which was liquid of salt bitter tears and a sorrowing heart comes as a sweet, cool and refreshing water to journeying souls to the Land of Spirits. And when the Aborigines visit this spot you will see tears trickling down their cheeks, the results of their thoughts of their Great Leader, the Messenger and Teacher of the Will of the Great Father of All.'

A train of tears also features in the Aboriginal songline associated with the journeys of the great ancestor hero Tjilbruke in the coastal country south of Adelaide. Tjilbruke himself was descended from Murray River people and he was said to have come down that mighty river with creator ancestor Ngurunderi.

Tjilbruke was in mourning for his sister's dead son Kulultuwi, who was killed for spearing an emu on another tribe's territory. He carried the body of his nephew down the coast starting from Brighton Beach and ending up at the tip of the Fleurieu Peninsula, at Cape Jervis, where he deposited the remains in a special cave.

Beside the kiosk at Brighton Beach, in Kingston Park, the spring is tucked away at the back of the beach. This is the place where Tjilbruke first mourned for his nephew. Where his tears fell, a permanent freshwater spring emerged in the Dreamtime, soaking down to the shore line. The site plays a significant role in the creation stories of the Kaurna people of the Adelaide plains.

At each place where the grieving man went on to stop and shed tears, a freshwater spring welled from the ground – at Hallet Cove, Port Noarlunga, Red Ochre Cove, Port Willunga, Sellicks Beach and Carrickalinga. So his tears can be seen today in the waters, a poignant reminder to his people to follow the rules of life and land.

Eventually Tjilbruke grew tired of life as a man and his spirit was transformed into a glossy ibis. His body was transformed into a rock outcrop of iron pyrites, for he was a master of fire making and his name means 'hidden fire'. Thus Tjilbruke's spirit lives on in the rock outcrop and in the ibis. The line of springs were used by the Kaurna people of the Adelaide Plains for their summer camps and the story of Tjilbruke's roundabout journey is partly a 'mind map' of clan territory and water sources, as told by the Putpunga people of the Rapid Bay and Ngarrinjeri Aboriginal groups.

One day in October 1991 the important, highly sacred spring at Brighton Beach was brutally bulldozed as part of a 'clean-up' by the local council, who were keen to extend the caravan park over it. While it appeared they had broken the law – *The Aboriginal Heritage Act* – council representatives insisted they did not have to consult Aboriginals or local residents about 'maintenance' of the important site, the *Guardian Messenger* reported on 24th October, 1990. Things are different these days and the spring is now allowed to flow unhindered in its small, swampy wilderness patch. On the hill above, a special stone monument, at the beginning of the 'Tjilbruke Dreaming Trail', explains some of the Aboriginal significance of the place.

For Catholics, the tears of Mother Mary embody the epitome of compassion. Mary has been manifesting at many places, especially at the holy wells and sacred springs of Ireland, and in other Catholic and also non-Catholic countries. And she is often seen to be in tears. Tears also spontaneously well from the faces of various Mary statues, such as the one known as the *Weeping Madonna* in Rockingham, Western Australia, as witnessed by the author in October 2006. The controversial tears that ooze down this Mary's plaster face are rose scented and oily. Many people who have witnessed the phenomenon have had a subsequent spiritual transformation in their lives.

Eye wells

The spot where water emerges from the bosom of the Earth is often called the 'eye' of the spring or well. The eye has to be kept clear for the waters to keep flowing.

Eye wells may also have an ancient connection to sun gods and goddesses, the eye often being representative of the sun, the 'all-seeing eye'. Sulis, a sun/water deity, rules the hot springs at Bath. The interplay and connection between fire/sun/light and water has long been a human fascination.

In Ireland we find places like Tobar na Sul which means 'Well of the Eye'. The sun and the eyes both radiate energy, while both water and eyes reflect light. The Irish sun god Lugh's grandfather Balor had a single, huge eye that could kill with a single glance. Lugh killed Balor with a slingshot through his eye, which fell to the ground. Where it sank down and made a hole this became Lough na Suil – the Lake of the Eye. In early Sanskrit writings Kapilla's Eye, or the Brahma Weapon, was a similar device that produced a shaft of light with great destructive power, writes author F W Holiday.

The well and eye connection even turns up even in the Bible, where Jesus tells a blind man to 'wash in the Pool of Siloe', after which his sight is restored.

As well as giving vision, a great many sacred pools and well waters have had an enduring reputation for curing eye problems. Some of the designated eye wells are found to have had little model eyes thrown in as votive offerings. This seems to be a widespread association. In Korea the goddess of water, Mulhalmoni, is invoked to cure eye disease or blindness; here coins are dropped into sacred springs and eyes are washed with the waters.

Well heads

Many European sacred wells, springs and rivers were also associated with human heads, in the form of stone carvings or real skulls thrown in as offerings. Sometimes the actual skulls of holy people were used to drink the waters from. Celtic people definitely had a stone head fetish and to them the head was the seat of power and wisdom. Often with Medusa-like features, snake-haired stone heads have been found at healing wells in the UK, France and Germany. Snake spirits have also been associated with water in the earliest spiritual traditions.

The head theme continues in folk tales that associate certain sacred springs with severed human heads. Certain sacred water sources were explained as gushing up spontaneously when someone holy was beheaded. Perhaps the most famous of these traditions is found at St Winefride's Well at Holywell, north Wales. This holy well maintains the longest unbroken tradition of pilgrimage in the UK, surviving against all odds since the 7th century. Ornate architecture and continuing patronage make this spring unique in the UK.

Oracle waters

The sacred water sources had many dimensions of import. Over the eons, tribal specialists who worked as intermediaries of the gods in the Otherworlds evolved a wealth of understanding of the wisdom of the waters. Long associated with oracular powers, European traditions of water as an oracular tool have survived through millennia, including centuries of suppression.

In Greece the renowned oracle centre at Delphi was once ruled by the sacred serpent spirit Python (originally of female gender) and dedicated to Python's mother Ge, or Gaia, the Earth Goddess. Its priestesses were known as Pythia. These women were consulted on some of the most important matters of state and the shrine's fame was at its peak around 2500 years ago. (Christianity had it closed by decree about AD385.)

A geological fault underlies the site, a factor that subtle energy researchers recognise as correlating with all sorts of phenomena. The presence of ethylene gas may be involved too. Several sacred springs used to arise nearby and were said to be the haunt of the three Muses.

The Wisdom of Water

To prepare for consultations, the Pythia would first bathe in the waters of the Castalian spring, then drink the sacred waters of the Cassotis spring to receive inspiration, before entering the temple. Here they would sit in a basement cell, on a sacred tripod stool, positioned over the point where the Earth's subtle vapours were released – possibly at an Earth fissure. They inhaled smoke of hallucogenic laurel leaves (and possibly chewed them) and would then go into a trance and make prophetic utterances, answering questions with cryptic replies. (Such a location, to a modern geomantic dowser, might be described as an Earth spirit serpent's 'breathing hole'.)

An even older oracle centre of great renown was at Dodona, in north-west Greece. The method of divination here was the inspired interpretation of the whisperings of the leaves of a sacred oak that grew above a sacred spring, or the gurgling sounds of the spring waters themselves.

Ireland has long been a stronghold for water oracles. Women known as *banfathi* would specialise in reading omens by viewing running water, which would be intuited by both its visible patterns and its sound. Sightings of the spirit woman *bean nighe,* the 'washer woman at the ford' were taken as portents of death, if she was seen washing blood-stained clothes.

Over in Scandinavia, where pagan traditions flourished longer than in much of Europe, a world serpent was said to live secured in the depths of the sea. In the Creation Song in the poem *Beowulf*, the cosmos is represented as a mighty World Tree (Yggdrasil) that lies in the centre of a round disc surrounded by an ocean. Beneath the tree is a sacred spring and a great serpent or dragon called Nidhogg lies coiled beneath its roots, curled around the circle.

By the Viking age this scenario had currency in pagan realms stretching from Germany to Siberia. It reflects the more ancient African representation of the universe as the horned serpent Aido Hwedo, known elsewhere as Ouroboros, that coils inside a great calabash biting its own tail – a symbol of eternity and the endless cycles of birth, death and regeneration.

Legend goes that Odin drank from the waters of the Well of Mimir that lies beneath the World Tree, in order to obtain great wisdom. He paid the price of one of his eyes. Mimir's guardian had been an ancient giant who was beheaded by the gods, after which Odin embalmed his head and was then able to consult with it whenever he needed urgent counsel.

In Ireland and the UK there are countless examples of water sources used for divination. Celtic tradition has it that fish may be the spirits of people transformed in legend, and sacred well fish and eels were once carefully protected by well guardians. The Scottish surname Dewar (Doire) reflects the ancient occupation of families with hereditary roles as oracular well guardians, Nigel Pennick writes. And next to St Peris's holy well (Ffynnon Beris) in Cornwall there was a cottage where the hereditary guardian once lived – a wise woman who tended two sacred trout that lived in the well. These were the agents for her oracular powers, and visitors who came to consult the woman paid a fee for her services.

At some wells pieces of old skulls, presumably belonging to important ancestors, local heroes and holy people, were used to drink the oracular waters from as part of the required ritual process.

Other waters communicated portents in their own right. In east Yorkshire, unusual spring fed streams – the 'gypseys' – flow intermittently over the chalk strata and have underground reservoirs. Sometimes called 'woe waters' when in full spate, they have historically appeared before major historical events. They gushed up just before the Restoration, the landing of William of Orange and the two World Wars, as well as before big localised sea storms and the fall of a great meteorite at Wold Newton in 1795. The largest one is the Gypsey Race, which is 22 miles long. It is also associated with the Rud Stone, a very tall and unique monolith still standing in an ancient churchyard in Rudstone.

The famous 'drumming well' at Oundle was said to 'drum' before disastrous events. It was recorded in 1691 to have drummed before the Scottish army invaded during the English Civil War, and again before the death of King Charles the second. In Northhamptonshire the Marvel Sike spring would start to run irregularly before disaster and St Helen's Well at Rushton Spencer, Staffordshire, would dry up – despite any wet weather – before local misfortune struck.

In north Scotland an oracular well on Maelrubha, an island in Loch Maree, is a small well next to a tree that is studded with coins, nails, screws and bits of iron, left in thanks as offerings. It is a pilgrimage place for the curing of mental disorders. Omens, also, were determined by the depth of the well – it was a good omen if the well was full.

St. Julian's Well in Wellow, Somerset has a history of visitations by a 'White Lady', who would manifest when a calamity was about to happen to the owners of the well. Visions of other such 'white ladies' at wells and springs have often been interpreted as 'Mother Mary, and have sometimes spawned pilgrimage traditions there. This happened most famously at Lourdes, in France.

Springs and wells 'predict' earthquakes and rain

Many of the holy wells of Europe are found in geologically faulted country. Changes to the depth of their waters has sometimes been recognised as a precursor to earthquakes occurring there and, in other areas, to volcanic eruptions. (Dowsers report big changes to Earth energy activity before seismic events too.) The traditions can vary. Gary R Varner writes that:
'Among the hill folk of the Ozarks it was believed, and may still be so, that the abrupt drop in the water level of springs or wells indicates that wet weather is imminent.'

Yet the opposite seems to be happening with a sign of imminent rain in Australian country folklore, told to me by central Victorian farmer John Murdoch, who said that:
'When dry springs start to moisten up or even start to flow into the creeks, that's said to herald the onset of good rains. I think it's a widespread observation'.

Perhaps the low pressure front is attracting water to the surface, just as an increased outgassing before a storm front arrives is observed to make Daylesford's mineral waters go cloudy. (Sea levels also respond to changes in air pressure, the Norwegian Hydrographic Service reports.)

Well dreaming at Glastonbury

In the mid eighteenth century at Glastonbury, Somerset, spring waters that emerge below the famous Tor there were channelled through the town for its water supply. One night in 1750 local yeoman Matthew Chancellor, who had suffered asthma for 30 years, had a violent asthma attack, then a vivid dream, which was written up in a local broadsheet. In the dream he was shown a spot that had 'some of the finest water I ever saw in my life'. He drank some, then stood up and saw a person who told him to drink the water every Sunday morning for seven weeks, and that where the water came from was out of Holy Ground. So he did as he was told and:

'drank it, returning God thanks, and so continued seven Sundays, and, by the blessing of God, recovered me of my disorder'.

Word spread rapidly and soon droves of the sick and infirm flocked there, from all parts of the country in the hope of receiving a cure. The water was bottled and sold in London. 'Many were healed and great numbers received benefit' it was reported, although there was a strong element of hype. The new spa industry took off with developments broadcast in influential newspapers by a local entrepreneur Anne Galloway, who was offering services to pilgrims. But despite supposed cures, the spa's bright promise never was fulfilled. Many people went home disappointed. Within a few years people were staying away in droves and the place was deserted.

Nothing more about the Glastonbury spa is found in writings after 1760, but the original spa building, a converted row of 17th-century cottages with a pump room and the stream running beneath it, was returned to it's former use when, in 1959, Wellesley Tudor Pole purchased the site to preserve it as a sacred site. It is now cared for by the Chalice Well Trust. Beautiful gardens of peace and serenity surround the well and stream, and the public can relax in its lovely atmosphere. The Trust's brief is to safeguard and protect the well and to provide facilities for meditation and spiritual awareness.

The chalybeate waters are stained red and are known for their high iron content; while another legendary stream nearby, emanating from underlying chalk strata, is called the White Stream. The well centre is on the corner of Chilkwell Street, probably a reference to the White Spring that seeps into the back of the Chalice Well garden.

The Chalice Trust states that the daily flow is 25,000 gallons per day and that:

'the healing power of the waters is not just in its mineral content, but in a subtle vibratory force that is released when the water leaves its subterranean home and interacts with the forces of the Earth, air and light above ground.'

There are several holy trees in the vicinity and the well is implicated in a tale from the King Arthur legends, where the Holy Grail of legend was 'deposited under the sacred hill from which the blood spring ran'. The stone shaft of the well is believed to be 700 years old and its cover, sporting an attractive *vesica pisces* symbol on the lid, was donated in 1919 by mystic archeologist Frederick Bligh Bond. Glastonbury has gone on to become a great spiritual and 'new-age' centre for the nation, thanks to its complex of holy centres and related mythos.

The Holy Grail connection makes this the most well-known British well in modern times, although it doesn't have the greatest of healing reputations, nor the long tradition of pilgrimage and cures, such as is found at St Winefrides Well. However, it is certainly the more pleasant well to visit, with its strong aura of peaceful meditation.

Continuing oracle traditions

To this day, the calm waters of the world are evocative of quiet reflection and meditative states, conducive to problem solving and enlightenment.

Quoting ancient texts, Rabbi Dennis says that:
'Water is a place of revelation ... In the mystical book The Vision of Ezekiel, the prophet received his prophetic insights by gazing into the reflection on the river Chabar'.

And in Tibet, each successive Dalai Lama has been selected partly by monks meditating on the calm waters of the sacred lake Llamo Lhatso, in the south. When visionary messages were seen as images appearing on the water's surface, the location of the re-incarnated Dalai Lama could be divined.

The Christian Church placed its own monopoly on such revelations. Papal and church decrees prohibiting the use of *frehtwellas* – springs used for divination – continued to be pronounced until quite recent times. Nonetheless, the indigenous, pagan practices continued to flourish in quiet country backwaters. The role of the wise woman at St Gulval's Well in Cornwall, for instance, was still active up until the 18th century.

Divination by the scrying of water (also known as hydromancy) is still practised by many of the world's folk, by those who are keen to discover and make good use of the powers of their psyche. One might sit by a body of water and watch the patterns playing on its surface and intuit their meaning. Or a spirit might manifest at the water and bring a message. Nostradamus is thought to have gazed into a small bowl of water as a scrying tool and received images of future events, many of which have since occurred.

By staring at waters, both still and moving, by watching water's reflections, by intuitive interpretations of the ripples of pebbles or patterns of oil thrown into water, and by similar such means, one is able to free up the intuitive mind and be more receptive to information from beyond.

References

Ginsburg, Rabbi Yitzchak, 'The Mystical Significance of the Hebrew Letters – M' at www.inner.org/hebleter/mem.htm

Archer, John, 'The Consciousness of Water', in *Green Connections*, Australia, February 1998.

Meehan, Cary, *Sacred Ireland*, Gothic Image, UK, 2004.

Rossbach, Sarah, *Feng Shui*, E P Dutton, USA, 1986.

Berndt Museum, University of WA, Nedlands, Perth. http://www.berndt.uwa.edu.au www.berndt.uwa.edu.au

Monaghan, Patricia, *The Book of Goddesses and Heroines*, Llewellyn, USA, 1990.

Forsyth, Chris, 'Drought: divining and pleading to the divine', *Weekend Australian*, 20–21st November, 1982.

Strong, Christopher, 'Irish Water', in *Dowsing Today*, vol. 40 no 281, British Society of Dowsers, September 2003.

Nelson, Robert A., *Air Wells, Fog Fences & Dew Ponds, Methods for Recovery of Atmospheric Humidity*, 2003, at: www.rexresearch.com/airwells/airwells.htm

Ringma, Lee and Frits, Homa Therapy Association of Australia, P.O. Box 68 Cessnock NSW 2325, ph 02 4998 1332 email – omshreedham@optusnet.com.au

Holiday, F. W., *Serpents of the Sky, Dragons of the Earth*, Horus House Press, USA, 1973.

Harte, Jeremy, 'The significance of 'holy', in *Well Researched*, Issue One (May 2000). The Wellsprings Fellowship, UK (on-line).
Haggert, Bridget, 'The Holy Wells of Ireland', in *The Wellspring Journal* (on-line).
Billingsley, John, in *Northern Earth*, no 76, winter 1998.
Chalice Well: www.chalicewell.org.uk
Body, Geoffrey and Gallop, Roy – *Healing Waters–the Mineral Springs and Small Spas of Somerset*, Fiducia Press, UK, 2005.
Pennick, N, *Sacred World of the Celts*, Inner Traditions, US, 1997.
Ross, Dr Anne, *Pagan Celtic Britain*, Constable, UK, 1967.
Cowan, David, *Leylines and Earth Energies*, UK.
Education Department South Australia, *The Kaurna – Aboriginal People of the Adelaide Plains*, 1989.
Rushe, Pauline, 'Sacred site dug up by Brighton Council' 17th October 1990 and 'Spring site 'cleanup' may be breach of heritage act', 24th October 1990, *Guardian Messenger*, South Australia.
Cairns, Hugh and Yidumduma Harney, Bill, *Dark Sparklers*, published by Hugh Cairns, 2003.
Screeton, Paul, *Quicksilver Heritage*, Abacus, UK, 1977.
Pennick, Nigel, *The Ancient Science of Geomancy*, Thames and Hudson, UK, 1979.
Unaipon, David, *Legendary Tales of the Australian Aborigines*, Melbourne University Press, 2001.
Varner, Gary R, *The Lore of Sacred Water*, 2003, at – www.druidnetwork.org/articles/garyvarner.html#f1
Bord, Janet, *Cures and Curses – Ritual and Cult at Holy Wells*, Heart of Albion Press, UK, 2006.
Monaghan, Patricia, *The Book of Goddesses and Heroines*, Llewellyn USA 1981.
McGrath, Sheena, *The Sun Goddess*, Blandford, UK, 1997.
Dennis, Rabbi Geoffrey W, *Water*, at www.pantheon.org/articles/w/water.html
Cunningham, Scott, *Earth Power*, Llewellyn, USA.
www.dalailama.com

2.4 Water Worship

Spirit of Water

While dismissed by many as merely products of imagination, the spirits of nature (also called devas) are actually an other-dimensional kingdom of life on Earth. Seen by clairvoyants in all sizes, shapes and colours, and detectable by dowsing as fields of devic energy, they exhibit vitality, emotion and intelligence. They interact with the flora and fauna, and often assist its development and evolution. (The totemic spirits of Aboriginal Australia are good examples of close bonds between the devas and human society.) They also reproduce, die (fading away, they are re-absorbed into the ether) and evolve themselves, just as does the rest of Nature.

Some are more at home in the Earth, or its water, air or fire; these devas are often referred to as elemental beings. But they don't necessarily always stay fixed to their elemental domains. Highly evolved devas can have influence over large regions, and harmony on Earth is all the better maintained when these local devas are happy and untroubled.

Indigenous peoples attributed sanctity to every water source and recognised its spirit denizens. They honoured its genius loci, usually a well nymph, and the spirits of water were generally revered above those of the other elements. In many cases it seems, human veneration of particularly important nature spirits has empowered and accelerated their development, and they have grown to enjoy elevated status as deities. Thus a region's devas became the gods, and – more so, the goddesses – of early tribal nations. Tribal identification with a tutelary deity, usually a goddess, was an early form of nationalism. The Irish had Eireann, and, in earlier times, Danu/Danaan of the Tuatha da Danaan tribe; while the British had Brit (Britannia).

The mythic gods provided prototypes, models of human behaviour for people to emulate, archetypes to inhabit the collective consciousness of mankind. Thus mother goddesses exemplified the maternal warmth, compassion, protectiveness and fortitude required to rear children. The gods influenced human society, while also being partly a product of it.

Thus these highly evolved spiritual beings, their forms somewhat fixed by human concepts, have evolved over time, reflecting the changing world around them.

The most highly evolved spiritual dimension of water – its heart and soul – may be referred to as the *Spirit of Water*. It is personified as a divine feminine being, a compassionate, creative, motherly figure, whose spiritual continuum could never be fully extinguished by Christianity and other imperialistic religions.

In the Western world She evolved by taking on the guise of the biblical Mary and now answers to that name, and to many other names too. Like Her many watery manifestations, this Great Mother, the Spirit of Water, may be found everywhere. But in waters that have been abused by mankind, her spirit grows weak.

Amphibious gods

In the creation myths of many ancient cultures, it was the gods and goddesses who emerged from the waters and brought with them the know-how to advance human culture. Watery or stormy deities were often the sacred ancestors, creators and law makers.

Often the gods were depicted in a half fish/half human form. Ea, for example, who was one of a trinity of Babylonian creator gods, was based in the primordial ocean surrounding, and supporting, the Earth. Part human, part fish (or sometimes a fish-tailed goat), he was a god of fresh waters, wisdom, magic and oracular powers.

Storm gods

Most of the leading Indo-European gods originally ruled the sky, from whence they bestowed rain and Earthly fertility, generally using lightning as their deadly weapon. Often weather/sky god and goddess pairs ruled the cosmic roost, their many children becoming the lesser deities. Though usually male, in Russia and Lithuania the thunder god is female, named Perperuna and Percuna Tete – the Mother of Thunder.

Baal, the Canaanite (western Semitic) god of death and rebirth, is a typical storm god. He resides on Sapan mountain and is depicted wielding thunderbolts. His voice is the thunder and he is associated with Earthly fertility. Aspects of the God of the Bible seem to have been borrowed from this thunder god.

Zeus/Jupiter became the leading sky god of the Greeks and Romans. Jupiter, in his thundering guise, was called Fulgar, the name later chosen for objects created by lightning. Virile Zeus, who sired a vast number of other deities, dispensed justice, and his thunderbolts were forged by the one-eyed Cyclopes. As Keraunos, Zeus was an ancient thunder god and Athenians prayed to him for rain during droughts, especially at Mount Lycaeus, his birthplace, where there was a sacred spring.

Zeus' totems are the eagle and the oak. Oak is known as an attractor of lightning. Oak trees like to grow over underground water streams, so it's not surprising that they get struck and that people traditionally never shelter under an oak in an electrical storm. At his famous oracle centre at Dodona, where a sacred spring arose from beneath an oak tree, Zeus' oracular messages were heard in the gurgling waters and rustling leaves.

The Baltic storm god Perkunas/Perkons, depicted as a red-bearded, axe-carrying man who drives his billy goat drawn chariot across the sky, dispenses justice and sends thunderbolts to evildoers. It was said that 'the first thunderbolt in spring purified the Earth and encouraged growth,' writes author Sheena McGrath. Perkunas was also a warrior and blacksmith. In Prussia Perkunas temples had eternal flames tended by 'vestal' virgins, and only oak wood was burned there.

Axes were popular with storm gods. Images of axes in the petroglyphs of Sweden and eastern Europe probably represent the storm god Thor. (The axe later evolved into a hammer, the god becoming a smith, a master of forging with fire and water.) One of the huge stones of Stonehenge in England has a faint axe carved onto it. And in Aboriginal northern Australia the Lightning Man, Namarakon, wears stone axes on his knees and elbows, and hurls them to make thunder.

We now know that lightning can make atmospheric nitrogen available to plants, so the ancient connection between lightning and fertility proves to be correct. If we add to the fact that lightning can actually create amino acids, the building blocks of life, it seems that our ancestors had already intuited the important functions that electrical storms and rain provided to the development and nourishment of life on Earth, if not to its very existence.

Water goddesses

The divinity of water in landscapes and oceans has nearly always been equated with the female principle. Goddesses ruled tribal territory. They were the chief divinity presiding over a watershed or bio-region, and all its waters were held particularly sacred to Her.

Over in India, in the earliest of sacred texts – the Vedas, dating from around the middle of the second millennium BCE – we find hymns of water goddess worship. A few are dedicated to The Waters, maternal goddesses who purify and nourish. The source of all the gods and universe, these mothers wash away guilt and bring health, wealth and immortality.

In Egypt *nt* – water- was probably the root word of several names for water spirits, including naiad, nixie and nymph. The Semitic word *mu* and Phoenician/Hebrew *mem* no doubt begat the names of the sea in French, la mer, and la mere/mother, plus water goddesses such as *Mary* and *Miriam* and the sea mother of the Finns and Saami people *Mere-Ama*. The *Morgan* are Celtic sea sprites and Morgan is also a sea goddess in Brittany, while in Ireland *Moruadh* is the green haired Irish sea maiden and *Muireartach*, the wild stormy, one-eyed sea goddess there. Mare is a female horse and water horse spirits are common in many global myths.

Of course not all water spirits are female and water can be yin or yang. Water goddesses often had consorts who ruled with them over various aspects of territory. And sometimes water goddesses were paired up with sun gods, the ultimate yang-yin connection. Such divinely dynamic duos are often associated with hot springs.

Goddesses have largely faded from the modern mind, but fish-tailed mermaids and the like fan the flames of their memory. Ireland's river goddesses prevail, as does the memory of *Cally Berry*, the 'water hag' of Northern Ireland, who has equivalents in the south and also in Scotland. This spirit of lakes protects them from being drained. She also controls the weather and created mountains; and when she appears as a crane with sticks in her beak, she forecasts storms.

Back in the Paleolithic and Neolithic eras, for over 20,000 years in Old Europe, it was mostly matrifocal, matrilineal cultures and their goddesses that ruled the divinity of nature and weather. If they were not depicted as sacred water snakes, the old goddesses were invariably shown as water birds in the ritual art forms that have been left as testimony.

Female deities were largely usurped with the Indo-European invasions of patriarchal tribes from the Russian steppes. A melange of the two spiritual polarities developed after 2500 BC and the classic cosmologies, in which gods raped and pillaged the goddesses, record those changing times in later, blended traditions.

Images of Celto-Roman mother goddesses have been found at sacred wells in Europe. *The Matres*, as the triple *Coventina*, are seen seated, holding a water jug in one hand and pouring out a stream of water from the other, in a carving at Coventina's well on Hadrian's Wall in England. Three goddesses are also depicted on a plaque discovered in the bathing complex at Bath. (They may be representing the triple-aspected nature of a single goddess, as maiden, mother and crone.)

Bath's resident deity Herself is depicted in a life-sized gilded bronze head, all that remains of Her cult statue. Her composite title *Sulis Minerva* reflects on the spiritual hybridisation that occurred in Britain's Roman times. Native goddess Sulis's name indicates an ancient sun goddess and also an indigenous well nymph.

The Romans added the name of Minerva, a goddess with several layers of guise, who seems to have had some of Greek *Athena*'s warlike aspects tacked onto her original attributes. But originally, Minerva was revered as an intellectual, 'wisdom incarnate in female form' and inventor of music, writes author Pat Monaghan. The Romans referred to Sulis as Minerva Medica, or *Healing Minerva*.

This is a reminder to us of how colonial cultures are often grafted onto ancient indigenous traditions (and vice versa) and their divinities recycled, repackaged and re-invented with fresh vigour down through the generations.

Queen of the South Seas

In the Indo-Pacific region, water deities are often sea-going. In Yogyakarta, the cultural capital of Indonesia, ancient animist traditions make this city beholden to a regal sea diety who carries the titles of *Loro Kidul* (Queen of the South Seas), *Kanjeng Ratu* Kidul (Supreme Southern Queen) and *Roro Kidul* (The Southern Maiden).

Manifesting as a beautiful goddess dressed in green, She controls the nearby sea and sea bed on the southern coast of Java. She is also the traditional protector of the once-powerful Mataram Dynasty and its current descendants, the sultans of Yogyakarta. The Queen is married spiritually to those descendants.

She is said to live under the ocean at Parangtritis, a beautiful black-sand beach village about 30 kilometres to the south of Yogya's city centre. When angered she can whip up tsumanis and sieismic mayhem, so the local people regularly appease Her with offerings to invite Her goodwill.

The Queen's waterfront shrine at Parangkusumo beach is close to where an original legendary meeting of the goddess and the local sultan once occurred, on two flat-topped pieces of volcanic rock. Here also is where the annual ceremony of 'Labuhan' is performed.

Each year, on the 30th day of the Javanese month of 'Rejeb', offerings are given to the goddess. These consist of food and clothing, plus all the Sultan's hair and fingernail cuttings from the previous year. The offerings are cast into the sea in the hope that the Sultan and the people of Yogyakarta enjoy continuous peace and prosperity.

In troubled times, sultans of Yogyakarta made their way through an underground network of caves to Parangtritis to seek out the Queen's advice. This last happened in the final days of Suharto's reign in early 1998, when the city's current leader made the journey, *The Australian* reported. The next day a million people gathered outside the palace and the sultan suggested to them that Suharto should step down, which he duly did, the next day.

Water shrines

Sacred springs offering life-sustaining waters were always protected from harm or honoured with special shrines. A sacred focal point for rising civilisations, often the water shrine was a central precinct of early settlements with ceremonial processional ways leading to it. Later, especially in the Middle Ages, healing springs sometimes had pilgrims' huts and sanctuaries built nearby, erected by the grateful.

An amazingly well preserved neolithic water shrine, dated at 4700 years old – the oldest of its kind – was found in 1995 in the Brecon Beacons in South Wales. The media reported how a local forest ranger made this rare discovery while digging a pond. Horizontally laid timbers formed a 9 metre (30 foot) walkway surrounding three sides of a stretch of water 18 metres (60 feet) by 13.5 metres (45 feet) long, with a wooden structure on the fourth, eastern side. In the centre of the pond stood a circle of a dozen wooden poles that may have risen to up to 6 metres (18 feet) above the water. These oak timbers may once have been connected together by horizontal wooden lintels. If so, the whole thing would have had a 'resemblance to Stonehenge'. The site has been excavated (some might say desecrated) by members of the Clwyd Powys Archeological Trust.

It was usual for Neolithic religious structures to be laid out in a circle or oval formed from wood or stone. Wet locations are known to have had great significance in those times, and many ritual deposits have been found in lakes and bogs. But this Welsh shrine is unique for its good state of preservation, reported *The Independent*, 7th December 1995.

The circular nature of ancient shrines suggest the idea of endless cycles, the circle of the seasons and the revolution of celestial bodies. That most well known of great circular temples, Stonehenge in England, turns out to be also connected to water – the nearby Avon River.

In 2006 excavations not far from Stonehenge found a henge – a circle of ditches and earthen banks – at Durrington Walls. This once enclosed concentric rings of huge timber posts, 'basically a wooden version of Stonehenge'. The new findings include a 'well-trodden avenue from Durrington Walls to the Avon River' and a village that may have housed Stonehenge's builders and/or semi-nomadic pilgrims of 4600 years ago. The ruins appear to be the largest Neolithic village ever found in Britain.

The roadway is paved with stone and leads to the Avon River, about 150 metres (160 yards) away. It is similar to a river road running from Stonehenge to the Avon.

Julian Thomas, a professor of archaeology at Manchester, thinks that the Durrington Walls timber temple was where the midwinter would have been celebrated and there is lots of evidence of feasting. Stonehenge appears to have been used for midsummer celebrations. These were more sedate affairs, which left no deposits of animal bones and broken pots. Thomas reported in February 2007 that:

'...*What this means is that Stonehenge, with its circles of stones, and Durrington Walls, with its circles of timbers, which are directly comparable, are actually linked together by this pair of avenues and the river. And this means that, effectively, they're one integrated structure.*'

Sacred rivers, lakes and springs

We know that sacred lakes and rivers were subjects of veneration since earliest times across Europe. This is evidenced by the numerous and sometimes magnificent artifacts that have been dredged up from underwater sediments. Items of weaponry and armour, as well as skulls and stone heads, were once ritually deposited in the sacred waters.

Displays at the London Museum reveal an unbroken practice from ancient times, starting with Stone Age tools, axes and blades and ending in Roman times with some very valuable and sophisticated weaponry and other items thrown into London's Thames River. Swords and lakes have mystical connections in the Arthurian legends, too.

Sun and water

There was an old belief, fairly universal, that the Sun travels through the Underworld at night, refreshing itself in the underground waters there. Some people sought healing at sacred springs by sleeping for a night beside them – perhaps expecting to gain the benefit from the nourishing powers of the underground Sun.

The Wardaman tribe of northern Australia also have such a tradition. They say the Sun sinks at dusk under the ground and goes into the subterranean watercourses there, which the Rainbow releases to the surface occasionally as springs. Wardaman describe the Sun as going down through a big tunnel into the watery Underworld, through a canopy of weeds which protect the underground waters and retain their warmth. The Sun returns to the surface each day in the east at dawn, all cleansed and refreshed. The Rainbow, that loveliest combination of sun and water, gets a mention too. A spring is the 'eyeball of the Rainbow', the Wardaman say.

On the other side of the world, at a French shrine at Mont Dol near St Malo, morning ceremonies at four sacred springs once celebrated the Sun's daily emergence from the Underworld. In Burgundy a major solar – water shrine had a sanctuary in the shape of a sun wheel. Beside the River Cure at Les Fontines Salees, this shrine was later developed and expanded by the Romans.

In pre-Christian Ireland the sun god *Lugh*, the 'shining one', was a 'perfect hero and saviour of his people', says Meehan. Likewise, *Aine* and *Grian*, landscape goddesses of Limerick, are also sun deities. They are possibly sister goddesses, Grian being the weak winter sun and Aine the bright summer sun. 'Aine' means bright or glowing. She is also known as *Anu*, and even *Danu*, in Kerry and elsewhere in the south-west.

Danu was the matriarchal chief goddess of the Tuatha da Danann tribe, whose arrival probably heralded the beginning of the Bronze Age in Ireland. The tribe were adept at working gold and bronze, much of which was emblazoned with sun motifs.

As sunny promoters of fertility, we are reminded of the goddess's fecundity by County Cork's Paps of Anu, two breast shaped hills, visual fonts of the goddess's goodness. Originally the harvest festival Lughnasa was held there. In Christian times it became a berry-picking festival held on 'St Latiaran's Sunday', around 25th July. A nearby well, named after the goddess *Red Claws* ('Crobh Dearg's Castle'), is the place where cattle were once herded to the water annually at May Eve, to ward off disease for the coming year.

Nearby at Cullen is St Latiaran's well with its attendant sacred white thorn tree and famous 'Curtsey stone', a heart shaped boulder honoured by women on the festival day, which is also 25th July. Latiaran had two sisters, these being Lassair ('flame') and Inghean Bhuidhe, the 'yellow-haired daughter'. Other older traditions of three local sacred women

– *Anu, Babdha* and *Macha*; and also of *Latiaran, Crobh Dearg* and *Gobnait* exist. When these three firy females disappeared from Earthly existence they left three sacred wells behind them, at Cullen, Cahercrovdarrig and Ballyvourney.

Bridgid was probably a later sun goddess and in Christian times St Bridget took over Her role. Bridget's name also refers to brightness. St Patrick replaced Lugh as a symbol of light and many famous annual pilgrimage celebrations of harvest, including the Lughnasa festival of around 1st August, were re-dedicated to Patrick at the holy wells.

In northern Europe there are many 'Helen's Wells' and these are associated with *Mother Hel/Dame Holle*. This goddess gave her name to 'holy' and 'healing' and Frau Holle was ubiquitous in the German-speaking world. She was a white lady who bathed at noon in fountains, from where spirits of children then emanated. She lived in mountain caves or wells and could be visited by diving into them. Sunshine was said to stream from her hair when it was combed, and rain would fall when she threw out washing water. She brought gifts of bounty and new technologies to the people, such as the invention of flax and spinning. Her feast day is the winter solstice.

In Germany one of the most important sacred sites is at Aachen. Here the imperial cathedral is located over the sacred healing well of Aquae Granni, which is dedicated to sun god *Grannos*, whom the Romans equated with *Apollo*.

Worship of the sun became demonised as Christianity tried to obtain supremacy. The sacred number of the sun – 666 – became transformed into a symbol of the Devil in the Bible.

Deities of hot springs

Hot springs were a favourite resting stop of the night sun, which was thought by the Celtic peoples to empower the healing waters. The geothermal spring waters at Bath were presided over by the sun goddess Sulis and another sun deity, a Grannos/Apollo snake-haired figure, is also depicted there.

Around Italy, *Burmanus/Borvo* was a deity of turbulent and hot waters and *Burmana/Damona*, was his female equivalent. Burmanus is depicted in a ceramic image with a horned serpent and a dolphin. He is also associated with a large sacred forest, the Lucas Burmani, around Cervo in Liguria.

River Goddesses

A belief in the holiness of all rivers is seen in countries like Britain, where six rivers have a name based on *'Dee'*, the word derived from deva, holy one. Resident river deities were usually female. The goddesses of the mighty rivers were the ruling deities of entire landscape drainage systems – queens of their watersheds. As well as being promoters of fertility of the bio-region, river goddesses were associated with healing, inspiration and prophecy.

Irish *Boinn* (or Boand), for example, is the goddess of the Boyne; *Belisama* (another whose name means 'bright one') presided over England's Mersey, *Sabrina* lived in the Severn and *Coventina* the Carrawburgh. In Scotland *Clota* ruled over the Clyde, Europe's Danube is ruled by the god *Danuvius*, while *Sequana* is goddess of the Seine. Valuable votive offerings were made to them, at their sources in particular, to ensure health and prosperity for all.

In Irish legend, the Shannon River is ruled by goddess *Sionnan* or *Sinend* (pronounced Shannon), who was the grand daughter of *Manannan Mac Lir*, god of the sea. At a shady pool called the Shannon Pot, in northwest County Cavan, the mortal Shannon went to eat of the 'forbidden fruit' of knowledge, the sacred salmon, giver of wisdom. But she paid a high price, for the enraged fish lashed its tail and the waters of the pool sprang up and overwhelmed her, their flow becoming the River Shannon. Shannon, meanwhile, was transformed into the river goddess.

Likewise Boinn (pronounced Boeen) was another woman of great curiosity, wife of the water god Neachtain, who, with his brothers, was the keeper of the forbidden magical Well of Segais, now called the Trinity Well. This is the traditional source of the Boyne River at Carbery, in County Kildare. (Carbery used to be called Sidh Neachtain – the fairy mound of Neachtain.)

Around this sacred well, in legend, nine magical hazelnut trees dropped their nuts of knowledge into the waters, to feed the salmon, the wisest being of all. But the well's treasure was well guarded, and when Boinn tasted its waters they, too, rose up in fury, pouring out a mighty flood and drowning Boinn while creating the River Boyne. It was said that anyone drinking from the well in June would become a poet. A pilgrimage still occurs at the river's source there on Trinity Sunday, the first Sunday in June.

Rivers were usually considered sacred from their source to the end and waterfalls were particularly revered. People would sit under waterfalls, as well as practise ritual immersion in pools, for restoration and renewal of health and spiritual wellbeing. The clergy of the early Celtic Christian Church continued this tradition and there are stories of them meditating while standing in rivers and holy wells.

Over in India, seven rivers enjoy the greatest sanctity and their ruling goddesses are the pre-eminent *Ganga*, *Yamuna* and *Sarasvati*. Sarasvati was the prototypical river devi of the Sarasvati River in north-western India, which has now vanished underground following a seismic disturbance. The Sarasvati was a river said to flow down from the celestial ocean, bringing multiple blessings with it and filtering through Shiva's dreadlocks as he sat meditating on the sacred Mount Kailash, dividing the waters into seven different streams that flowed down into the Earth.

Goddess Ganga is celebrated at the confluence of three rivers at Allahabad, where a huge annual pilgrimage takes place. Later hymns identify Sarasvati with the goddess *Vag*, the goddess of speech. Vag created the universe and brought language and poetic vision. She is called the 'mother who gives birth by naming things' (a role seen in many other creator deities the world over). Sarasvati also took on such attributes over time and she is now associated with learning, poetry, music and culture.

In Nigeria *Oshun* is the pre-eminent river goddess of fertility and healing for the Yoruba tribe. Her husband is *Shango*. During the times of slavery Her cult spread to Middle and South America. Oshun is the 'owner of the sweet waters', the goddess of love, sexuality, beauty, pleasure, happiness and diplomacy. She is also responsible for fertility, love and divination. While generally she is a great giver, when she is angry it can be very difficult to calm her down.

Magic and healing waters

Sacred springs are found in all sorts of locations, from rocky hillsides to mountain tops and passes, seashores and bogs. Often there is a sacred tree growing next to the sacred well. Traditional in Celtic countries, the ash tree is frequently associated with sacred springs.

In times when water storage was difficult and water cleansing treatments unknown, spring waters provided the purest water available and thus quickly established curative and restorative, as well as magical, reputations. The wise, tribal peasant doctors knew which medicinal herbs to add to spring waters to effect their folk remedies. Such folk traditions were kept secret for fear of persecution during the times of the Inquisition, but they never disappeared completely.

A 16th century book *The Life of St Columcille* mentions many healing wells and tells of the crowds of pilgrims who came in the mid 14th century to St Mullin's Well to pray and walk in the stream waters there. 'Some came through devotion, the majority through fear of the plague,' a commentator noted. Some wells were called Tobar na Plaighe – the Well of the Plague.

The sacred waters were visited often and for a range of purposes. They were taken to cure various ills, by both man and beast, as well as for psychological problems and infertility. There were also wells used for magic (surviving as Wishing Wells), as well as for cursing, for affecting the weather, for deflecting storms and for attracting rain.

Fishermen on the Isle of Man would visit the Wells of the North or South Wind and give offerings in the hope of procuring favourable winds at sea. Before heading out, they would ritually throw handfuls of water into the air towards the preferred direction of the wind.

Such water magic isn't always available every day and at Loch na Naire in Sutherland, Scotland, the water was said to only have magical and medicinal qualities between midnight and 1 am on Lammas Day and May Day.

Some healing wells were reputed to be helpful in cases of infertility and King James II and his second queen visited St Winefrede's Well at

The Wisdom of Water

Holywell in Wales in 1686, in the hope of procuring a child. Other wells were renowned for curing particular minor conditions such as warts, rheumatism, whooping cough, skin diseases and backache. Iron rich waters such as the 'chalybeate springs' of the Chalice Well in Glastonbury were often associated with healing. And some wells had a reputation for curing love sickness.

It is curious that a great many healing waters were associated with curing eye problems. 'People suffered badly from vitamin A deficiency during mediaeval times, and the primary symptom of that is sore eyes,' explains well researcher Katy Gordan.

As a cure for mental illness, people used to be thrown into sacred waters for a bit of shock therapy. A Tobar na n-Gealt – the Well of the Lunatics, in County Kerry – people drank the clear waters and ate the watercress that grew there. In Scotland the mentally ill were immersed three times in the waters of Loch Maree, or towed around the island by boat. It didn't always work though.

It was always customary at the healing wells to give offerings in gratitude for services rendered. This usually took the form of a piece of rag, a bent pin, nail, needles, coins, flowers, food or drink. Often those seeking cures would tie a rag to 'clootie' trees or shrubs. These can still be seen today in Ireland covered with colourful strips of rag. The rags were traditionally a piece of the clothing of those seeking cures and symbolised the health problem that they were leaving behind. They are meant to rot away as the ailment ceased. Unfortunately the clootie trees of today can look very messy, with synthetic rag materials refusing to break down!

At St Lassairs Well in Roscommon I saw several pens, presumably left to invoke good results in examinations, as thanks for academic assistance. But the commonest votive offerings in Irish wells today are holy pictures, rosaries and candles. And at St Winefride's Well at Holywell in north Wales, around 50,000 candles a year are burned in a century-old tradition.

At some Roman-Celtic water shrines there was once a brisk trade in models of body parts made from oak, ivory, and copper alloy, or recycled from old statues. These were purchased and flung into the waters in seeking a cure for that body part. At Bath and other Roman water shrines, people would write their problems on lead scrolls and throw them into the water.

To this day, in some quiet country corners, whole communities get involved in decorating their local sacred springs at annual well dressings. The tradition is strong in Derbyshire, England, where the decorations – arrangements of flowers and greenery – have a Victorian flavour. (The custom may well have been a Victorian era revival.) Traditions of water rituals are also found at Droitwich, Worcester, where dressings of St Richard's well occurred up until the Civil War. When the practice ceased, the waters dried up. The custom was then revived and the water came back!

Healing properties

Modern scientific investigations into holy or healing waters of Europe, measuring their electrical fields, have found significant differences between them and the ordinary, non-sacred springs or well waters tested.

Traditional healing waters have also been found to be rich in minerals such as germanium, which helps maintain high oxygen levels in water and is highly therapeutic, writes author Charlie Ryrie. The miraculous healing of Catherine Latapie on 1st March 1858 was the first recorded cure at France's Lourdes. These days, 2500 unexplained healings later, Lourdes water is found to be high in germanium.

As well as minerals and gases in water, the effect of the prayers of pilgrims, their state of meditation and their thankfulness at sacred sources must all be absorbed energetically by the sacred waters and help to provide the overall curative effects and lovely atmosphere one finds at such places.

Balinese water goddess

On the fabulous volcanic island of Bali, in Indonesia, an intricate sharing system provides irrigation water for the fertile rice paddies upon which the island depends. To be successful, rice growing requires careful timing of planting, crop rotation and the efficient and timely use of large quantities of water.

The Balinese revere the island's waters as a divine gift from their water goddess *Dewi Danu*, who oversees the complex management of an elaborate network of irrigation channels administered from temples controlled by priests. Bali has many famous water temples that are located beside its lakes, irrigation canals and reservoirs. The priests of these temples are the managers of the water and arbitrators of water disputes.

Pura Ulun Danu Bratan in Bedugul is the most important and well known of the water temples. It presides over a picturesque volcanic crater lake nearly a mile (1.6 kilometre) above sea level, where the air is much cooler and lush pine forests thrive. This temple, consecrated in 1663, is the seat of Dewi Danu and here She is attended by her 24 priests. The high priest, Her human representation, with the title Jero Gde, was chosen in childhood by the goddess to serve until his death.

The temple controls about half of Bali's agricultural land with a highly productive system that involves an intricate complex of canals, weirs, tunnels and ditches connected to 74 square miles of paddy fields. Festivals take place here as well, and the temple is an important social hub.

The Balinese have relied on their ritual-based system of irrigation for at least one thousand years. The temples are strategically placed at each level or diversion of the irrigation path. Every field has a temple in a corner too, and here farmers show gratitude for the rice crop with prayers and offerings.

Rice farmers are organized into co-ops called subaks. Subak leaders ascend to the temple where prayers, offerings and meetings take place to affect a good crop and ensure that subaks far away downstream will have enough water to sustain their paddies. With age-old crop rotation patterns, the rice farmers experience few problems with pests. And all is controlled by the beneficent Goddess of the Lake.

Water and spirit in the Bible

The biblical Holy Spirit is traditionally feminine, says feminist writer Barbara Walker. Her emblem is the white dove and, as *ruach* in the Bible and Hebrew traditions, She hovers or broods over the waters as a creative mist. Sometimes described as weather or wind, a storm or a whirlwind, the word *ruach* combines meanings of breath, mind and spirit, and is similar to the Greek *pneuma*. Sometimes *ruach* is identified as 'air, gas from the womb'.

In the Bible we also read of the pool of Bethesda, where 'an angel went down at a certain season into the pool and troubled the water; whosoever then first after the troubling of the water stepped in was made whole of whatsoever disease he had.'

Elsewhere in the Bible, other examples of magic, healing and divinely energised waters are found – no doubt a reflection of older, animist traditions. Holy water may be found in churches, but certainly no-one's church has a monopoly over sacred waters!

Angel of water

A similar angelic being, reminiscent of the biblical *ruach*, has been observed by clairvoyant friends on my own property in Victoria. Dove shaped, and several metres across, it was hovering over the waters of my newly built dam (which had, amazingly, filled in its first fortnight and during a drought). Its head was seen to be focusing downwards as it hovered above, and from it a beam of energy projected down into the water. I guess this was its way of 'troubling the waters', of charging them up with its power and knowledge. The dam waters there maintain good quality.

Water and purity

In 2006 when I visited St Lassair's well in Roscommon, I met an Irish lady and her son. Hunched and diminutive, the 88-yeat-old told me of a successful back cure she had received there some 70 years before. 'Mammy', as her son called her, had followed an age old tradition of drinking the restorative waters, then crawling on hands and knees beneath a nearby stone table, to cure her backache. The stone table is reminiscent of a dolmen in miniature. Two vertical slabs are surmounted by a horizontal top flagstone. On its top sits a beautiful and intriguing water washed spherical stone about 30 centimetres (1 foot) across. Such stones are known as a 'serpent's egg'.

People have been gathering at this spring to celebrate harvest and fertility long before Christian times, as the presence of the fertility inducing serpent's eggs would suggest. It has probably been a sacred site since Neolithic times. Old Mammy said she knew nothing about the serpent stone, which has miraculously survived Church suppression. But she did tell an interesting tale, pointing up the hill with crooked fingers:

'The spring used to come out further up the hill there ... But some fellows were carrying on up there next to the source. They were cursing and blaspheming ... and so it moved! And now it's down here.'

The connection between water and purity has been a long one. Remember that the word 'well' also means 'good'! That water refuses to suffer pollution – whether physical, mental or verbal – is seen in the many legends of wells moving or disappearing when insulted in some way.

Yet water also has an amazing ability to cleanse and purify itself, to cancel out patterns of pollution and regenerate itself, with the assistance of nature's little helpers.

Lustration

Universally, and since earliest times, water has been regarded as a symbolic vehicle for the purification and spiritual renewal of mankind. Rites of ritual immersion are called 'lustration', 'lustre' suggesting the radiant or luminous brightness of the illumination of wisdom. Ancient rites of initiation often involved lustration and some pagan water rituals were written about in the Old Testament.

In Japan's indigenous, animist religion of Shinto the worship of kamis (the devas) always begins with the all-important act of purification with water. Inside sacred shrines, troughs are kept for ritual washing. During Suigyo/water austerities waterfalls are stood under for spiritual purification.

In Islam, Muslims must be ritually pure before prayer. For this purpose some mosques have a courtyard with a central pool of clear water for ablutions, but in most mosques this is done outside the walls. Fountains, symbolic of purity, are also sometimes found in mosques.

In Judaism, ritual washing is done to restore or maintain a state of ritual purity. Usually just the hands, or the hands and the feet are washed; sometimes total immersion is required. This was ordained to be done in 'living water', that is, the sea, a river, a spring or in a mikva (ritual bath) of 'living water'.

Rituals of purification with water were later established in most Christian churches and sects. Jesus was baptised by John the Baptist in the sacred River Jordan and later Jesus commanded his disciples to be also thus baptised. These days baptism is simply used as a public declaration of a person's belief and faith and as a sign of welcome into the Church. The Catholic Church, however, believes that a real change occurs at baptism – that the 'stain of original sin is actually removed from the individual'.

In biblical Jerusalem, the waters of the Shiloah pool, which come from the nearby Gihon spring, were used in purification rituals, such as before visits to the Temple Mount. A 2000-year-old road adjacent to the Shiloah pool, uncovered in 2006, was once used by the masses during Jewish pilgrimage holidays in the Second Temple Period. Pilgrims immersed themselves in the Shiloah pool before entering the Temple Mount, say archeologists. A plastered hewn-stone mikva from the same era was unearthed at subsequent excavations. It once received water channelled from Solomon's Pools near Bethlehem, located several miles south of Jerusalem.

Churches and wells

Despite great efforts made to stamp out animist traditions in Europe and elsewhere, many ancient water rituals have continued to flourish in disguised forms under the banner of Christianity. Reverence for water was deeply ingrained. In 1704 an act of parliament forbade the practice of 'well worship' and well pilgrimages in Ireland, with a penalty of a public whipping or ten shilling fine. In 1770 a curate of Bromfield in England decreed all pagan ceremonies at his local Hellywell spring banned.

But well and fountain worship continued to prove difficult to eradicate. It had been popular throughout the Middle Ages, and survived the Reformation. In Ireland, after the terrible famine of the mid-19th century, when the population was devastated, much well wisdom and patronage went 'down the gurgler'.

A great majority of the sacred wells, in Ireland at least, are found located close by the ruins of medieval parish churches or the like, and as these were the central focus of authority it is not surprising that water was needed close by. Ley line pioneer researcher Alfred Watkins, in the UK, also rediscovered that there were alignments of all sorts of ancient structures associated with the locations of wells.

Alignments ('ley lines'), he found, typically run between sighting points such as prominent hills, or 'sometimes from a holy well to a hill and vice-versa', writes author Danny Sullivan. Even the design of cities like London was influenced by the presence of the sacred water sources, as author David Furlong has discovered.

Some churches were even built right over an old sacred well and early Celtic churches would have used them for baptism, until the Roman church replaced them with a font inside the building. Old churches with a crypt or grotto containing a subterranean spring can still be found here and there. Examples are at Holybourne Church in Hampshire and St Bride's in Fleet Street, London (the latter probably once a spring dedicated to the goddess Bride/Bridget).

Holy water is not the same water as holy well water. Holy water has been blessed by priests for use in certain rites, such as at the Easter Vigil, at blessings, dedications, exorcisms and burials, etc. It may have come from an ordinary tap. The custom of sprinkling people with water at mass began in the 9th century. At this time 'stoups', basins for holy water from which people could sprinkle themselves on entering a church, were in common use.

While originally underground waters were thought to be empowered by the sun at night, later, in Christian times, water was sometimes charged up energetically by contact with holy relics of saints. These would often be dipped into the water for a homoeopathic or placebo effect. The latest known cure from such a tradition was in 1963, when holy relics were dipped into the fountain at Le Grand Andeley in Normandy during the vigil and feast of St Clothilde, and sick people were immersed in the water afterwards. These days that tradition is largely forgotten and the fountain has degenerated into a simple wishing well.

The practice harks back to earlier times, when folk healers world wide knew to put special stones such as quartz crystal, the stone of the sun, and gems and medicinal herbs into sacred well waters to impart healing energies. The tradition continues today in various forms, such as the modern healing modalities of homoeopathy and the use of gem and flower essences and elixirs.

Well pilgrimages

In ancient Ireland there were at least 3000 sacred wells, with a long and vibrant pilgrimage tradition. Popular times to visit them were at the special times of the year, at *Imbolc* on 1st February, *Beltaine* on 1st May, *Lughnasa* on 1st August and *Samhain* on 1st November. These turning-points of the Celtic year were deemed to be times when the gates of the Otherworld were opened – a time for visions and divine inspiration. Later, in Christian times, the dates and names of the festivals were changed a bit, but the practise of honouring the waters continued.

The tradition of holy pilgrimage remained strong, mainly in the Catholic countries until the Middle Ages, and in England well until late medieval times. It enjoyed a revival of interest there in the 17th and 18th centuries as well. After the Reformation, the pilgrimage tradition was mostly continued only in Ireland, however, and it is still practiced there today, but on a smaller scale.

The Irish pilgrimage, or 'turas', focuses on a particular complex of sacred sites and objects, and usually features a holy well. A variety of 'stations' are visited on a certain day or time of year, and these are 'rounded' or circumambulated, while prayers are chanted. Pilgrims 'doing the rounds', sometimes on their bare knees, drink from the well or touch or kiss certain sacred stones, according to a set pattern. The term 'pattern day' refers to this old conglomerate tradition.

The Irish and British well pilgrimages became important social occasions, especially on Sundays and the feast days of the Christian saints associated with them. Every district in Ireland had its saint, whose name and legends were attached to various topographical features. The vernacular saint has been described as the genius loci – 'the spirit of place itself, more accessible and homelier than remote divinities', writes author Stiofan O' Cadhla. The saints have taken the place of more archaic heroes, both men and women. Many wells are dedicated, or indeed 'owned' by the local saint, *St Bridget*'s *Wells* are many, and there are many St Bridgets! The ancient goddess Bridgid (Bride in the UK) ruled healing, medicine, smithcraft and inspired poetry, as well as carrying a famous cauldron, that may well been the prototype for the Holy Grail.

Bridget's name refers to brightness. On her feast day of the 1st February, the traditional first day of spring, Bridget Crosses are made for her in the shape of quartered circles – sun symbols. As well as being associated with sacred wells – and perhaps the ancient tradition of the underground sun – St Bridget's name and attributes are like those of Sulis, and so she may well be the successor of the vernacular well nymphs and goddesses.

But this isn't always the case, author Patrick Logan discovered, recalling the St Bridget pilgrimage held in his home town in County Leitrim. Logan would join pilgrims in reciting the rosary, while approaching the local cemetery. Here they walked three times around an old ash tree and later knelt beside a carved stone supposed to represent the head of St Bridget. In the 1970s this stone was excavated and cleaned and found to have a beard!

Many wells were associated with St Patrick and legends tell how the saint

Photo: Holy well near Sligo, Ireland

displaced 'druids', who were formally in charge of the wells. Patrick used the well waters to baptise converts. Sometimes a well miraculously appeared, in answer to his command. Or, in the case of the well that is now hidden below the edge of Nassau Street in central Dublin, beneath the entrance to Trinity College, it was said to be brackish until St Patrick's magical prayers made its waters sweet and pure again. St Patrick's feast day is 17th March and this is when his wells are visited for 'patterns'.

Other wells are visited on Trinity Sunday, or Easter, on May Eve or May Day, the beginning of summer. The midsummer feast of 24th June, now the feast of St John the Baptist, is another popular feast day in Ireland. The old midsummer rituals began at sunset of the day before, up on hill tops where fires were lit. People waited around the wells for the midnight hour, when tradition had it that the water 'boiled up' and for the next hour was capable of curing any ill. Most of the subsequent Christian festivals that occur between 22nd July and 15th August were designed to replace the Lughnasa festivals that originally celebrated harvest and the light god Lugh.

The last Sunday in July, for instance, is the pattern day at Tobernalt, near Sligo, and mass is said there twice that day. There are fourteen 'praying stations' around this beautiful sacred grove and well area, some of which are merely piles of small stones. Each is numbered with Roman numerals, to represent the 'stations of the cross' in a Catholic church, which the ritual replicates.

In the past, pattern days often became racy affairs after dark, when piety was done with. There was lots of dancing, feasting and love-making. In his 1840 book *Holy Wells of Ireland* (which is available as a free download from the internet) Father Charles O'Connor, enquiring about the well at Beer Crowcombe in Somerset and the early 20th-century gatherings there was told that:

'*The well had the property of bubbling or moving, even steaming on some well known days of the year, notably Whit Sunday and during the fortnight following. At this time, while waiting the moving of the waters, a rendezvous of the youth and maidens from the villages round took place... There seems to have been a good deal of horse play around the well.*'

The annual patterns had become important events of social interaction, but often the heady mix of alcohol and testosterone saw them ending with fights, sometimes of a factional nature. A Popery Act of 1704 banned the 'riotous and unlawful assembling together of many thousands of papists to said wells'. The festivities were worrying the authorities, who were really waging a war of cultural imperialism and genocide, and for whom the pattern represented a 'sign of Irish anarchy, lawlessness and irreligion', says author O' Cadhla.

When O'Connor enquired of people why they attended well pilgrimages, a very old man told him that the people continue the tradition because
'...their ancestors always did it, that it was a preservative against Geasu-Draoidacht, i.e. the sorceries of Druids; that their cattle were preserved by it, from infections and disorders; that the daoini maethe, i.e. the fairies, were kept in good humour by it; and so thoroughly persuaded were they of the sanctity of those pagan practices that they would travel bareheaded and barefooted from ten to twenty miles for the purpose of crawling on their knees round these wells, and upright stones and oak trees westward as the sun travels, some three times, some six, some nine, and so on, in uneven numbers until their voluntary penances were completely fulfilled.'

The Water Goddess today

In the current era we still encounter the Spirit of Water as the great mother of all. In one notable example she has evolved into the composite being known as *Mother Mary*, with many millions of devotees worldwide. Mother Mary often manifests to little children or clairvoyants as a gentle and compassionate blue-white or silver being or energy field, and she is often encountered at holy wells or springs such as Lourdes.

The biblical Mary has little to do with this being who has evolved over the last several thousand years. The name 'Mary' is a word for the sea, the marine environment, where mariculture is done. Mother Mary is sometimes referred to as the 'Star of the Sea'. An earlier goddess – *Isis*, she of a thousand names – was called *Mere* in her sea goddess form; *Hathor* also had this same title. *Miriam* is a derivative name.

In Ireland well pilgrimage traditions go in and out of fashion, but often it is Mary who provides a continuing thread to modern times. A holy well was revealed in a vision around 1985 near a Cistercian monastery at Mount Mellleray in County Waterford, Ireland. A statue of Mary had been put at the site in a verdant valley, after which 'one night a glowing light lit up the stream from the spring,' writes author Janet Bord. There were several visions of Mary and messages were given to pray more and to recognise that the water was blessed. A shelter for visitors and a 'Lourdes Grotto' have since been erected by pilgrims.

In Australia the Anglican Church or Our Lady of Yankalilla was the site for a series of Mary visions and phenomena in the 1990s, which caused it to become a pilgrimage centre for people from far and wide, many of whom receiving healing. (I have written about this in greater detail in *Divining Earth Spirit*.) Holy water is drawn from beneath the church and bottled there.

The church site itself is located over the top of an Aboriginal sacred site. This was also the scene of a terrible massacre of the local Aboriginal tribe, when people of all ages were killed. An odd site for a church to be placed, perhaps. But when I visited and tuned in to the energies of place there, there was no sense of any 'bad vibes' from the massacre. The effect of thousands of pilgrims' prayers and Mary's compassionate energy had healed the pain. The loving-wisdom of the Spirit of the Waters has prevailed!

References

Walker, Barbara, *The Woman's Encyclopaedia of Myths and Secrets*, Harper & Row USA, 1983.
Monaghan, Patricia, *The Book of Goddesses and Heroines*, Llewellyn, UK, 1981.
Jordan, Michael, *Encyclopaedia of Gods*, Kyle Cathie, UK, 2002.
Gimbutas, Marija, *The Goddesses and Gods of Old Europe – Myth and Cult Images 6500 – 3500BC*, University of California Press, USA, 1982.
'The Goddess and the Computer' BBC Channel 4, Jan. 1989 via *MacUser*, January 1989.
O' Cadhla, Stiofan, *The Holy Well Tradition*, Four Courts Press, Ireland, 2002.
Logan, P, *The Holy Wells of Ireland*, Colin Smythe, 1980, UK.

Meehan, Cary, *The Traveller's Guide to Sacred Ireland*, Gothic Image, UK, 2002.
'*Java's South Sea Goddess*', ABC News Online 17th May, 2006
Lie-Birchall, Barrie, 'Parangtritis–A Beach Not Too Far', at: www.bootsnall.com/travelstories/asia/sep02beach.shtml
Fitzpatrick, Stephen, 'Disaster strikes at heart of Java identity', *The Australian*, 29th May 2006.
Jordan, Katy, 'Wells in Depth', *The Wellspring*, May 2000 .
Pennick, Nigel, *The Ancient Science of Geomancy*, Thames and Hudson, UK.
Screeton, Paul, *Quicksilver Heritage*, Abacus, UK, 1977.
Sullivan, Danny, *Ley Lines – the Greatest Landscape Mystery*, Green Magic, UK, 2004.
Wilford, John Noble, 'Unearthed: 4600-year-old village of Stonehenge builders', 1st February 2007, *The Age*.
Jordan, Katy, 'Wells In Depth' – Issue One (May 2000), *The Wellspring*.
Haggerty, Bridget, 'The Holy Wells of Ireland', *The Wellspring*, online journal.
'*Christianity has baptism – but Paganism had it first*' at www.medmalexperts.com/POCM/pagan_origins_baptism.html
Lefkovits, Etgar, '*Road to Temple Mount uncovered*' www.medarch.blogspot.com/2007/01/temple-aqueduct-and-ritual-bath.html
McGrath, Sheena, *The Sun Goddess*, Blandford, UK, 1997.
Bord, Janet, *Cures and Curses – Ritual and Cult at Holy Wells*, Heart of Albion Press, UK, 2006.
Isaacs, Jennifer, *Australian Dreaming*, Lansdowne Press, Australia, 1980.

Part Three:
Waters of the sky

3.1 Droughts in Australia

Diary of a drought

While rain is abundant in Australia's tropical north, in much of the south it has been ten years since 'normal' amounts of rain have fallen. Following this dry spell, the drought pinched very hard across southern Australia in 2006 and the stark and painful reality of it even became evident to city folk!

2006 was a year of climatic extremes, of see-sawing highs and lows and record-breaking temperatures. From early winter to late spring Victoria experienced an onslaught of crippling frosts, some down to over minus 5 degrees Celsius, and, in mid October, a widespread 40-odd degree heatwave struck. It was Australia's hottest spring on record, as well as one of the driest springtimes in many parts. Melbourne had one of its driest recorded years, with only half the average rainfall and by April 2007 every region in Victoria was deemed officially drought declared.

South-east Queensland, where the population has been increasing exponentially, was also suffering its worst drought on record. Two thirsty coal-fired power plants were only able operate at 80 percent capacity due to dwindling water supplies, Tim Flannery pointed out on 23rd May 2007 (*Lateline*, ABC TV). 2006 also saw Perth record its driest ever year, having only 477 millimetres of its 800 millimetre average; while down in the south-west they had the driest ever winter.

Victoria, December 2006
Like many other regions, people in rural Victoria had been forced to cart water in for household and stock supplies for some time. In some cases whole towns had been trucking water in. It's an expensive and time consuming activity. The ground was hard underfoot. It was about as dry as a landscape you can get without totally withering. And when it's this dry there's always the prospect of bush fires, here in Victoria, the 'bushfire capital of the world'. And it happened.

Tinder dry forests in the north east exploded into flame after dry lightning storms set them ablaze. Ten years of under average rainfall plus summer heat and evaporation had sucked away all moisture and the fires, fanned by hot winds, ravaged nearly one million hectares. Firefighters poured in from all over the state and interstate to fight the raging fires, which were now threatening to engulf towns. Things couldn't get much worse.

Just when greater disaster seemed inevitable, the weather took a miraculous turn around and mercy struck! In the midst of mid-summer, a cold front with icy rain came unexpectedly through, quenching the flames of the massive wild fire fronts of the north-east. It even snowed on Christmas day in the Alps, quenching fires there, while hail battered fruit crops elsewhere.

January 2007
Fires can re-start from smouldering stumps when suitable conditions return. Fire-fighters can't rest long when they have to fight potential fire with fire. During cooler spells a massive program of back burning had been undertaken. In January 2007 the skies were still grey and smoky, an ominous blanket that had been hanging around for weeks. The smell of burnt life was a constant reminder of danger and anxiety levels were high.

Some of the many fire fronts continued to rage. Exhausted firefighters (mostly volunteers) were still at work on day 45 of the fire emergency. New Zealanders and Canadians had flown in to help out, with a rotating force of over 3000 firefighters needed daily on the ground to battle and mop up after fires. There had been casualties, one death and dozens of houses lost.

17th January
The usual peak of midsummer heat had come to Victoria. Yesterday there was a state-wide Armageddon and the carnage of the bushfires even came to town! A day of stifling heatwave, fires were raging on many fronts in the north-east, with more being started by dry lightning strikes. Vast swathes of forests, and now farmland and houses, were burning. The premier Steve Bracks, back from holidays early, called it a day of historically unprecedented fires, noting that the fire season was still only young.

Worse was to come! The main power supply coming in from New South Wales, to the north, was under threat. An official on the ABC radio news around 6 am that morning had warned of possible trouble. But no-one else seemed to be expecting the state-wide blackout that affected millions of people from 4 pm up until, at latest, midnight that night.

Enveloped in dense smoke, the heavy concentration of carbon, as ash in the air, caused electricity to jump across to adjacent parallel power lines, tripping the whole system. Instantly the system went into automatic shedding and redistribution of power. A quarter to a third of the state, in varying areas, was blacked out. It was the power company's worst nightmare.

Much of Melbourne's peak-hour traffic was thrown in chaos, with traffic lights out and trains having massive delays. People melted in the heatwave as they waited long hours for transport home and had to contend with no power to run air-conditioners. The power company's emergency phone number merely gave a recorded message that there was an emergency situation and to hang up now! There had been no warnings to the public.

More hot weather was forecast for later in the week but, miraculously, instead – seasonally unusual good rain started falling over the state and more danger was averted. It had been a roller–coaster of a week!

Droughts, politics and déjà vu

As predicted by climate change models, Australia's already extreme climate has been heating up. To prepare for the weather of the future we need a good understanding of what is 'normal' rainfall. Climate change demands forward thinking, but this is wasted without the lens of the past. Oddly, the benefits of hindsight seem to have been unrealised when it comes to drought in Australia. This land of recurring droughts has been in denial. The image of parched landscapes with desperate inhabitants has historically not been a look that the government has wanted its people, as well as the rest of the world, to see. Temporary-sounding terms such as 'water shortage' or 'water famine' were once adopted to lull the population. A 1919 film, *The Breaking of the Drought*, was banned from overseas distribution because it showed realistic drought scenes in western New South Wales.

But climatic extremes and empty rivers have also become useful for pushing political agendas and power grabs. Damming rivers, for one, is good for politicians to gain "brownie points". The damming of some of Australia's last wild rivers is currently underway. Already south-east Queensland's Burnett River has had the devastation of the Paradise Dam – a desecration of wilderness and one of the last homes of the endangered lungfish, an ancient creature surviving from dinosaur times. Now the proposal for the Mary River dam has torn apart the thirsty community and people are facing the loss of their farms, plus more lungfish habitat will be lost, if it does go through. But dams don't guarantee any more rainfall. So why are they being pushed through? The World Wildlife Fund, quoted in *The Central & Northern Burnett Times*, 27th July, 2006, believes that:
'The decision to build the Burnett River Dam appears to be mainly politically motivated.'

The 'Federation Drought' of 1895–1902 may have been as severe as, or even worse than the current drought. (Really good rains didn't actually come until 1906.) By mid 1896 the Murray River was at an 'unprecedented low level' and irrigation supplies were stopped. Later, in the 1914 drought, the Murray went down to 1.7 percent of flow. 'The Murray', it was then announced, 'has ceased to be a river.' By April 1915 the Glenelg, Wimmera, Moorabool and Campaspe Rivers in Victoria had also ceased to flow. The Darling River at Bourke was virtually dry, the wheat crop lost.

Earlier droughts

Droughts had already been experienced in the earliest days of white colonisation in New South Wales and Victoria and were first recorded around 1838-1840. In western Victoria the first white settlers found Lake Colac almost dry in the late 1830s and Aboriginal people spoke of being able to walk across it in times past. Someone crossing the Pyrenees mountains found the three rivers there had largely dried up. After a promise of 'Australia Felix' the new settlers were very disappointed. Aboriginals living around the almost dry Lake Burrumbeet in the late 1830s and early 1840s blamed the drought on the invaders, saying:
'When the white people come – the water goes away.'

Another classic Australian drought, in South Australia 1864-1865, came after farmers and pastoralists had been moving a bit too deeply into the dry interior, on the strength of several decades of reasonable rainfall from the 1820s. This drought prompted the drawing of 'Goyder's Line' of rainfall, which indicated the limits of where country could be successfully cropped. A powerful surveyor-general for the colony, Goyder drew his line largely along vegetation markers, having noted the demarcation zones where Mallee trees gave way to salt bush, in drier country.

But when that drought was followed by seasons of plentiful rain, a huge land rush to the north of South Australia ensued. Memory was short and under pressure from would-be farmers, the government relented in 1874 and allowed more wheat farming north of Goyder's line. For a while South Australia was the booming wheat bowl of the country. And the idea that "rain follows the plough" and the Christian ethic that God would reward hard working farmers with rain, seemed to be well borne out by the seasons of abundance. But nowadays most of the ill-planned towns and homesteads in those parts lie in ruins.

Since the Federation Drought, rainfall patterns typically show a see-saw of highs and lows, which average out together to make – average rainfall! Southern Australia is really no stranger to suffering long years of drought, for they have occurred with great frequency, with widespread droughts experienced in 1911-1915, 1927-1929, 1943-1946, 1959, 1961, 1967, 1976-1977, 1982-1983 and 2002-03. In fact Australia's rainfall is said to be the most variable in the world, with records showing that for only 20 years out of the last 100 has the nation been drought free.

What of El Niño?

As for El Niño (dry) and La Niña (wet) climate events, this is based on oscillating Pacific Ocean patterns of atmospheric pressure and warm ocean water circulation that influence Australia's rainfall, mainly in the north and eastern seaboard. Rainfall is scant when the Southern Oscillation Index (SOI) is negative (but not always, as we shall see) during an El Niño event. The wet/dry periods tend to alternate about every three to seven years. However, they aren't always that regular, with the time from one event to the next varying from one to ten years. They can also cluster together and these have historically been ruinous to some early cultures.

Extreme El Niño events also affect weather patterns right around the globe. A bad series of El Niños, confirmed by Antarctic ice cores from around 4000 years ago, reveal the probable cause of the end of the great Sumerian city of Ur – as simply a sustained spell of wild wet weather. Its relatively rapid demise is recorded in clay tablets, but only recently has the weather connection been confirmed, a BBC TV documentary (*The Life and Times of El Niño*, 2005) informs us. Perhaps a long lasting El Niño event is the origin of the many archetypal legends of the Great Flood, of rain 'falling for 40 days'?

El Niño seems to also have caused the downfall of the Moche of northern Peru, a sophisticated civilisation referred to as South America's equivalent to classical Greece. Characterised by its colourful pottery and enormous mud pyramids, recent archeological examinations have found a possible explanation for the Moche's rapid decline around 1300 – 1400 years ago. The environment there is normally arid, yet the mud pyramids are half melted away. Just as the civilisation was blossoming, the climate suddenly changed. Furious storm seasons prevailed for around 30 years. This was followed by 30 years of drought and the abandoned cities are filled with desert sands that blew in after the flooding times.

Confirmation of the theory came from ice cores that went back to that era, sampled from an Andean mountain glacier. After these severe El Niño/La Niña events the Moche stopped making pyramids and were more concerned with survival. New walled towns sprang up and there was a preoccupation with self-defence. Civil warfare was no doubt spurred by the meagre resources left available in the devastated environment. Moche descendants still live in the region, making the same pottery, scratching a living in a much degraded environment.

A strong El Niño occurred in Australia in 1982–1983, corresponding with one of the severe drought periods previously mentioned. El Niño returned in 1997-1998, a period when the long and continuing spell of below average rains began in Victoria and elsewhere. But it was not as bad as the media hype had suggested. Quite the opposite. There was even flooding in many parts. Peter Hastings of the Queensland University of Technology has concluded that:

'The hazard of the El Niño Southern Oscillation and its impacts are complex and uncertain. There are temporal and geographical dimensions to its degrees of influence and impact, and of course, inherent uncertainties in its behaviours and effects. In fact, no two El Niño events are exactly the same.'

And La Niña?
Back in March 2006 forecasters were tentatively giving out the hope of a La Niña event for later in the year. Weak indications of a wet winter and spring for eastern Australia were given, but 'nine of the twelve models the weather bureau had analysed predicted neutral conditions. Two predicted a weak La Nina, while one pointed to warm conditions,' reported *The Weekly Times,* 18th October 2006. However they included a rider, with Meteorological Bureau analyst David Jones saying:
   'The Pacific has been behaving rather strangely.'

A slightly positive SOI, the possible early hallmarks of a La Niña, brought reasonable rains to the dry north. But in the south-east over 2006 the drought became entrenched. On 1st December 2006 the Bureau of Meteorology were hedging their bets and coyly suggested:
   'We have been in the grips of an El Niño [and] there's as much chance of above-average rainfall as there is for below-average rainfall.'

None of south-eastern Australia's above-average rain in 2007 had been foreseen by the Bureau. La Niña is not always a sure thing. Since 1878 34 La Niña events have been studied and it was found that only about one half of them brought above average rains, while about a third brought average and the rest below-average falls, in areas north of the Great Dividing Range, *The Weekly Times* reported (18th April 2007). The Bureau is looking to upgrade some of it's predictive models.

Climate change impacts

Many scientists think that climate change is making El Niño events more frequent and extreme. For the planetary systems of heating and cooling are disrupted and now act more chaotically than ever. Extra heating of the central Pacific Ocean has upset the usual balance and the whole world's weather can be widely affected by this. Certainly the weather has been wild all over!

In the west of Australia rainfall has been declining over the last 30 years. Tim Flannery says that this is a result of rising sea surface temperatures in the Indian Ocean. And it's also the reason for the prolonged drought, lasting for over the last 40 years, in Africa's Sahel (the sub-Saharan region extending from the Atlantic to Sudan). The rise of atmospheric greenhouse gases is the reason for the Indian Ocean warming up, he says, noting that:

'Climate models indicate that about half the [Australian rainfall] decline results from global warming, which has pushed the temperate weather zone southward. The Australian climatologist David Karoly thinks the other half results from destruction of the ozone layer, which has cooled the stratosphere over the Antarctic, thus hastening the circulation of cold air around the Pole and drawing the southern rainfall zone even further southwards.'

The likely impacts of climate change were announced recently by scientists of the International Panel on Climate Change (*The Weekly Times*, 18th April 2007). The projected scenario for Australia and the world is fairly grim but it's not all bad. Droughts will be getting worse and water availability could drop by 30 percent, they think. Higher temperatures in cooler areas will enhance crop growing, but not in the tropical north; while in southern Australia there will be more heavy and damaging summer rain events, amidst a generally lower rainfall. So, if models are correct, we can expect more desertification, occasional crop losses from unseasonal downpours, and more bushfires and floods.

What about climate cycles?

David Bellamy is a climate change sceptic. The last ice age ended some 13,000 years ago, he wrote in the UK *Daily Mail*, 9th July 2004, because a natural global warming process began. He says it is the long term 'Milankovitch Cycles,' which dictate the climate and it is:

'...nothing to do with the burning of fossil fuels... [A recent scientific study has shown that] ...increases in temperature are responsible for increases in atmospheric carbon dioxide levels, not the other way around ... But this sort of evidence is ignored. The real truth is that the main greenhouse gas – the one that has the most direct effect on land temperature – is water vapour, 99 percent of which is entirely natural. If all the water vapour was removed from the atmosphere, the temperature would fall by 33 degrees Celsius. But, remove all the carbon dioxide and the temperature might fall by just 0.3 percent'.

Other evidence points to the fact that solar radiation has been increasing of late, says UK researcher Dr Solanki. He explains that:

'The Sun is in a changed state. It is brighter than it was a few hundred

years ago and this brightening started relatively recently – in the last 100 to 150 years ... The increased solar brightness over the past 20 years has not been enough to cause the observed climate changes but ... the impact of more intense sunshine on the ozone layer and on cloud cover could be affecting the climate more than the sunlight itself.'

In agreement with Solanki, Professor Lance Endersbee, in his controversial recent book *Voyage of Discovery*, challenges conventional theories about global warming. He questions the assumption that an increase in atmospheric carbon dioxide has been the main cause of the apparent rise in the Earth's surface temperature, saying that:

'The major causes of variations in climate are variations in heat flow from the interior of the Earth, and variations in solar and cosmic radiation. These factors are not included in the present computer models of climate change.' (Monash University News, 23rd November 2005).

Soil and rainfall

Ross Hercott of Pyramid Hill is a Victorian farmer who has eliminated soil salinity on his farm by not using conventional fertilisers and by special deep ripping, with his self-designed Ecoplow. According to Hercott rainfall has been declining in Australia because of the way we have been farming. Acid fertilisers are especially bad, he says, because they dry out the soil.

In *The Weekly Times*, 13th April 2006, he said that:
'We stopped using acid fertiliser in 1984 and the first thing you notice is the amount of production you get without it, then you start to notice you get your rainfall back.

'I believe that the use of acid fertiliser is the main ingredient that has changed the world's rainfall so radically. By the simple fact that acid absorbs water, acid dehydrates water.'

Droughts are followed by floods

> **Tiddalick's Tale**
>
> In an Aboriginal legend from Victoria, Tiddalick was the frog who greedily drank every drop of water from all the waterways and ponds. The terrible drought that resulted caused much hardship for all nature's creatures. What to do?
>
> Eventually an eel tickled Tiddalick and made him laugh, causing all the waters to pour back out from his mouth – making a huge flood, which replenished all the land. And then all was well.

Confirming Tiddalick's tale, time and time again, drought is broken in Australia by often perilous flooding rains. Since 2006 ended, some spectacular rain events have occurred in eastern Australia and the centre. In Alice Springs, the Todd River, normally just sand, became a frothy, raging torrent, reaching its highest level in six years! Northern South Australia got a drenching on 20th January 2007, with up to 150 millimetres falling over 48 hours, and record-breaking flooding rain fell in western New South Wales and outback Queensland, with Bedourie, deemed the driest place in Queensland, and Birdsville flooded in! Townsville in north Queensland then had its heaviest rain in seven years and the region was flooding. The Burdekin Dam overflowed. Flooding rains poured down from Mackay up to Cairns. Unfortunately, the parched south east of Queensland and down the coast to Sydney still largely missed out. But by June heavy rains started to fall in those regions and a massive storm raged for three days, causing havoc and flooding of the Hunter River on the New South Wales central coast.

Drought and fire

When soil, pastures and forests are tinder dry, a bush fire is always a threat in the backs of peoples minds. Not only do bush fires raze buildings and kill people and animals, the ash can pollute air and water catchments and short out electricity supplies.

Devastating bush fires, often sparked by dry lightning storms, are highly prevalent in Victoria. In the state's recorded history, *Royal Auto* reported, the worst bushfire seasons were in 1851, 1898, 1926, 1932, 1939 (the shocking Black Friday fires), 1943, 1944, 1952, 1962, 1965, 1968, 1969,

1972, 1977, 1981, 1983 (the deadly 'Ash Wednesday' fires), 1985, 1997, 1998, 2002, 2003 (with a million hectares of alpine forests burnt out). The 2006 fire season, which started in December, razed dozens of homes and killed one firefighter. The total fire toll has been high: over 352 people dead and millions of dead stock, as well as wildlife; over 3600 homes lost; and something like over 8 million hectares burnt out, all up.

The January 2007 rains brought relief from the sweltering heat and dampened many fires, but were at first considered a nuisance for weary firefighters who were attempting to back burn before hot weather returned. The fire season was far from over! Flare-ups could happen anytime. But the cooler weather did continue and eventually, after 60 days of fire fighting, the fire risk in Gippsland was declared over. On 20th February it was announced that all the firefighters had finally gone home. Nineteen thousand had been involved, including 270 from overseas. It had been a mammoth effort. The firefighters were the heroes of the day.

Smoke and air pollution reduce rainfall

Smoke from bushfires has another insidious effect. It damages rain bearing clouds and reduces rainfall. This has been a dire effect on a large scale in areas downwind from massive forest fires in Indonesia, for example. And air pollution from industry and cars (diesel burning in particular) is not only unhealthy to breathe in, but also affects clouds and reduces rainfall, David Leaman points out. (A pulp mill currently proposed for Tasmania could also potentially reduce rainfall in the region, he suggests.)

The effects of such 'negative cloud seeding' by air pollution were described in a paper presented by Israeli scientist Daniel Rosenfeld at the Australian Cloud Seeding Research Symposium held in Melbourne in May 2007 (and it's freely available on-line).

Drought and frost

Sufficient levels of moisture in soil can effectively act as an anti-freeze for plants. Crop lands that are poor in organic matter, humus or clay are typically low in moisture. So when a drought drags on, the incidence of heavy frost generally escalates, making more nightmares for farmers and gardeners! Deciduous plants often die from the combination of low moisture and frost striking at the vulnerable stage when they are putting out new spring leaves.

Dust storms

Moisture, as a component of humus, also helps to stop precious top soil from being blown away in dust storms. Tilling of farm land in the Mallee during droughts has occasionally resulted in big storms blowing away massive amounts of top soil and dumping it in the Tasman Sea or on New Zealand, as happened in Melbourne on 8th February 1983. That severe dust storm was an eerie and haunting event, embedded in the psyche of the inhabitants. I happened to be there for a meeting of environmental groups, on the day that turned to night. The massive dust cloud was estimated to cover 500 kilometres, from Mildura to Melbourne, 150 kilometres from east to west and between 350 metres and 3500 metres high. A wild black storm front, it also cut power supplies and damaged homes.

Six months before then, farmers and the soil conservation authority were well aware of the drought's effect on soil erosion. It was estimated, for instance, that 80 percent of cropping paddocks west of Mildura and around Hopetoun were experiencing 'significant drift' of topsoil, Keating writes.

Sinking and shrinking

As the drought has lengthened many home-owners have found themselves watching large gaps develop between de-hydrated timbers and cracks growing in their walls, caused by the intensely dry air and dry soil affecting foundations. Homes built over reactive clays can be particularly affected. This appears to be the second time this problem has become noticeably severe.

During the 1982-1983 drought around 4 percent of Melbourne's buildings were estimated to be affected this way, with sometimes extensive structural damage, a side-effect of drought not previously noted. The Royal Australian Institute of Architects advise people to water around the perimeter of buildings to help stop the cracking if this occurs.

Drought and wildlife

Prolonged drought has dire effects on wetland wildlife. Fish in East Gippsland were reported to have been attacked by parasites who favour the warm, shallow waters of their depleted rivers.

Waterbirds have suffered an enormous decline of 82 percent between 1983-2004 (*The Age*, 5th June 2007). In other places, animals are on the move, searching for feed. Mobs of starving kangaroos and emus overcome their usual shyness and come to town, desperate for a feed, and more aggressive than usual. Breeding ceases.

Toxic soup

Low levels in reservoirs and waterways plus hot weather can create a toxic soup when nutrients have leached off the land and become concentrated in the water. Algal blooms, including toxic blue-green algae, can be rampant and many water supply reservoirs were subject to outbreaks over the summer of 2006-2007. Warm waters, low flows, slightly alkaline pH and clear, calm water all foster the growth of algae.

Blue-green algae can be invisible to the naked eye, but a large concentration will appear as a slick or scum. Nasty, musty odours are a warning sign. Physical contact with blue-green algae can lead to skin rashes, sore eyes, ears or nose and, if it is swallowed, gastroenteritis, nausea and vomiting can occur. Pumps and plumbing can get blocked with it too.

If a dam becomes infested remove any debris that has washed into the water then the conventional method is to apply doses of gypsum and alum and, when things have settled, a dose of chlorine to sterilise (*The Weekly Times*, 18th April, 2007).

There are also non-chemical treatments you can try. The application of blue metal (crushed basalt rock), when broadcast into the water, can be a cure. Zeolite is another useful rock dust, that's particularly good at mopping up nitrogenous material. And some organic farmers prevent blue-green algae by placing bales of organically grown barley hay around the edges of their dams, where water flows in.

Water Theft

Another problem of drought is the rise in thefts of water, water tanks and equipment. *The Weekly Times* reported a string of thefts across the drought stricken Wimmera (10th Jan, 2007), with one incidence where a Country Fire Authority tanker had been drained dry.

It could be worse, I suppose. In other parts of the world people are at war, killing each other over the sharing of drinking water. Take Israel and Palestine, for example. It's dry country and water is scarce. International law states that most of the water sources in the area are international resources and must be shared equitably. But this is not happening. Disputed Palestinian territories, where as much as 40 percent of Israel's water is sourced from (such as the Eastern Aquifer in the West Bank and the Gaza Aquifer) were forcibly taken by the Israelis in the 1967 War. Palestinians have rejected past peace proposals from Israel because, amongst other things, it didn't give them control of water resources within their territory.

How much water on Earth?

Rob Gourlay:

'It is estimated that ocean water comprises 97.3 percent of all of the Earth's water, and 2.1 percent of water is tied up in ice and glaciers. Fresh water is 0.6 percent of the total water budget, with the bulk of it tied up in groundwater. The atmosphere has 0.0001 percent or 13,000 cubic kilometres of this fresh water and along with rivers and lakes make up 0.05 percent of fresh water aboveground. Therefore, groundwater is overwhelmingly the greater source of fresh water for life.

'Precipitation provides 4000 cubic kilometres annually, albeit with spatial and temporal variability, however most people live in areas of reasonable rainfall (i.e. coastal areas in Australia). Rivers carry 1953 cubic kilometres and groundwater is recharged with 432 cubic kilometres of rainfall. People use 690 cubic kilometres of surface water and 393 cubic kilometres of groundwater, therefore the surplus water available for life is massive; provided some of this water can be effectively produced (stored and protected from wastage), harvested and recycled'.

Where does the water go?

The twenty first century heralded an era of global water shortages. Over one-third of the world's population currently experiences serious water problems and polluted water makes over one billion people sick each year. Between 1900 and 1995 global water consumption rose six-fold and that was at over double the rate of population growth! *(UNESCO Sources* No. 84, November 1996*).* Asia's so-called 'green revolution', which began in the 1960s, saw the amount of agricultural land under irrigation tripling and now many parts of the continent have reached the limits of water supply. The breadbaskets of India and China are both facing severe water shortages, a conference was told in New Delhi in January 2007.

Australians are some of the biggest water guzzlers on the planet, yet we live on its driest continent. Australians have the third highest per capita water use in the world, after the USA and Canada. It was estimated that in 1996-1997 Australia used 24,000 gigalitres of water – 75 percent of it for irrigation, 20 percent for urban domestic and industrial use, and 5 percent for rural domestic and stock use. Our water use has been rising dramatically. Between 1984 and 1997 irrigation increased by 76 percent, total urban use rose 55 percent, although rural (non-irrigation) use fell by 2 percent.

Melbourne has had a long history of problematic water supply. But thanks to the creation of reservoirs in the 1860s exotic, water drenched gardens became the rage and Victoria became known as the 'Garden State'. In 1983 38 percent of all household water in Melbourne was used on gardens. Currently Melbourne's total water for residential use accounts for 60 percent, while commercial and industrial use accounts for around 28 percent. Gardens still account for 35 percent of domestic water use.

An early 2007 report pointed out that some country Victorians use twice as much water as their city counterparts. Householders in Mildura and Swan Hill were each using an average 552 kilolitres in 2005-2006, when there were no water restrictions. (As of December 2006, stage one restrictions were invoked.) The average use for the rest of regional Victoria at that time was 240 kilolitres per household and just 191 kilolitres in metropolitan Melbourne, where tough restrictions applied.

References

Hunt, Peter, 'A History of Wet and Dry,' *The Weekly Times*, 18th October 2006.
Hastings, Peter, '*El Niño: Hype and Hope the Australian Way*', Royal Geographical Society of Queensland, at www.rgsq.gil.com.au/elninoC.htm
Bellamy David, 'Global Warming? What a load of poppycock!' *Daily Mail*, 9th July, 2004.
Leidig, Michael and Nikkhah, Roya, '*The truth about global warming – it's the Sun that's to blame*', at: www.telegraph.co.uk/connected/main.jhtml?xml=/connected/2004/07/19/ecnsun18.xml
'The Big Burns: Victoria's Worst Fires,' *Royal Auto*, RACV, December 2006.
Keating, Jenny, '*The Drought Walked Through*', Department of Water Resources, Victoria, 1992.
'La Niña Teases Experts', *The Weekly Times*, 29th March 2006.
McCallin, Jessica, '*Blood and Water*', Grist Magazine, 26th February 2002 via www.ifamericansknew.org/new.html
'Rainfall dives as mercury soars' *The Advertiser*, Victoria, 1st December 2006.
'*Water Recycling in Australia*', Australian Academy of Technological Sciences and Engineering, 2004.
Flannery, Tim, '*We Are the Weather Makers – the history and future impact of climate change*', Text Publishing, 2005, Australia.
Cloud Seeding Conference: www.bom.gov.au/bmrc/basic/events/cloudseeding/CS_Booklet.pdf
World water use: www.epa.nsw.gov.au/stormwater/hsieteachguide/waterpoln.htm
Randeep Ramesh, World is running out of water, says UN adviser in New Delhi, 22nd January, 2007, *Guardian Unlimited*, at: www.environment.guardian.co.uk/water/story/0,,1996211,00.html

3.2 Weather prediction by nature

For all the sophisticated equipment and computer modelling used by meteorological experts, we still seem to be completely taken by surprise by many, sometimes major, weather events that occur. Perhaps we ought to take more notice of nature. People living close to nature have their own methods of weather prediction that are often much more accurate, especially on a medium to long term basis, than high-tech methods.

It's all in the flowering

Some people have refined their weather forecasting observations to a fine and accurate art. Retired farmer Mervyn Obst, of Jeparit in Victorians northern Wimmera, is one. Obst was instructed by his old uncle in how to correctly predict rainfall many months in advance. His uncle had relied on the method since the early 1900s, when there was no other source of information. (Although he may well have been informed by local Wotjabaluk Aboriginal people.)

For 40 years Mervyn was taught to understand the patterns of flowering of the local black box eucalyptus trees. He went on to correctly predict the late breaking rains of spring 2006. Earlier, in late 2005, he saw that box flowering was light and short lived and gave the thumbs down for decent rain in 2006. Correct again.

But he had some good news for us also. In March 2006 he reported prolific bud growth on box trees around Jeparit, suggesting massive flowering in the summer and good rains for 2007, around the northern Wimmera, at least. Trees heavily loaded with summer blossoms were also apparent in my region of central Victoria. And as it has panned out – some unusually heavy falls have come to the west and centre of the state in the first half of 2007, which is normally the dry half of the year in the driest half of the state.

Obst had forecast good rains in May 2007 – and possibly an even earlier autumn break for the farmers of the west, if thunderstorms persisted. He had never seen the local box trees flowering so prolifically. 2007 should be a wet year for the northern Wimmera and maybe even too wet for some, he reckoned, in *The Weekly Times* of 10th January. Just ten days later his predictions had proved, once again, correct.

20th January 2007. Over the last few days the north of South Australia was mopping up after a 1 in 50 year rain event that saw Hawker and Port Pirie flooded. Rain then swept through western Victoria, delivered soaking, record-breaking events over several days, with 114 millimetres falling in Casterton on just one day – double their previous January record, the Bureau of Metereology said. The centre of the state received a bit less, while the east remained drought-stricken. The box flower prediction for Jeparit was looking good.

More was yet to come! Obst had suggested that the autumn break, which the farmers were anxiously waiting for, would come in May or earlier, if storms kept up. The ideal window of opportunity for winter crop sowing is if rains fall between Anzac Day (25th April) and the 1st of May. But April had been desperately dry. Things were not looking good. Many farmers would be ruined if they couldn't get a crop in, after the failure of last year's winter rains. Many reservoirs were just puddles, virtually empty. Stock feed had almost run out.

Then, out of the blue, around 27th April the heavens mercifully opened up! The Mallee and Wimmera got a great drenching, while the Grampians area received 63mm in just one day. A normal autumn break would be good if just 25 millimetres fell and this was the best in the west for decades, with some areas receiving more rain in the first few months of 2007 than they'd had in the whole year before. Many farmers had never before seen so much rain in one time. And with elevated soil and air temperatures (up by around two degrees Celsius) – it was magic for pasture growth and seed germination! A glorious green mantle covered the land.

Animal behaviour

Despite the many heavy downpours, there was no need to worry about the nesting Black Swans on the lakes of Victoria's west. In December 2006 swans had been observed building nests much higher than usual, an aquaintance was told by her brother on 7th January. He said that:

'The "old-wives-tales" talk at Hamilton is that the drought will break in the next few months, because the swans at the local lake have built their nests one metre (3 feet) above their normal level, and they've laid five instead of three eggs in them. Also, the gum-trees are coming out of a sort of shut-down, a dormant period of minimal growth. "Floods" is the whisper'.

As 127millimetres (5 inches) or so of rain went on to fall over that region on 20th January, it's a lovely illustration of the amazing intuition exhibited by nature. And it's often a hot topic in country folklore.

Dogs and cats
Many people would be familiar with the observed behaviour of dogs and cats apparently sensing the approach of storms long before we do. Cats often obsessively lick their paws and ears, for example.

Ants
Ant behaviour is another classic sign. When ants become agitated and carry around little white eggs and pupae on their backs, moving them to safe ground, or building the rims around their nests high, they probably are preparing for wetter times. A blogger on an ABC website wrote that:
 'I have noticed that the ants tend to react more to the changes in barometric pressure than the actual rain. Most rain events follow a drop in barometric pressure.'

Romance for 'roos
Rabbits and kangaroos start to get frisky and snakes slithery when good rain is imminent. Kangaroos are known for their ability to rapidly multiply after a drought. *The Weekly Times* reported on Valentines Day 14th February 2007 that kangaroos had been observed to be very amorous around the Grampians Mountains at that time.

Subsequently the region enjoyed above average rains. It came just as Grampians water storages, which provide a lifeline to the Wimmera and Mallee, had plummeted to 3.6 percent full, an official on ABC radio said on 18th May. Enough lead up rain had already saturated the catchment and now the critically low storage reservoirs had a chance to receive some run-off. Already, over the previous few days, storage levels had risen to 4.1 percent and it was looking like a crisis had been averted. The Grampians received 242 millimetres of rain that May, which was well above average!

Kookaburras have a laugh
Kookaburras also forecast rain. They'll laugh hysterically before it rains, but it must be before 11 am, I've been told.

Busy bees
Tree blossoms must have also been abundant in the Central Highlands region, keeping the bees very busy. Someone in Trentham told me that:
'My bee keeper friend told me that it would be a wet year in 2007, he could tell from the bees.'

Termites
Termites (white ants) will typically swarm before humid or wet weather. A website blogger also described how:
'... grandfather had a termite mound he used to watch (for 15 years at least), and when a fresh addition appeared on the mound he reckoned that major rainfall (25 millimetres +) was about a week away. He was pretty accurate'.

Nature has incredible foresight. It has this ability thanks to the amazing sensory faculties and intelligence of life, whereby plants and animals sense and interpret Earth's energetic and atmospheric changes, in order to foretell coming weather, or dangerous events, and survive it. Dowsing is an extension of this form of intuition. People the world over have also developed a wealth of know-how about natural weather prediction. It is often just a case of enquiring from the old timers in a district, if you don't know what to look out for yourself. The tell-tale signs will be different in different places.

Aboriginal weather wisdom

Ancient Australian Aboriginal knowledge is very well informed in this regard. The value of their indigenous weather lore has been recognised by Australia's Bureau of Meteorology, who say on their website that:
'a significant amount of the weather culture of the Aboriginals of central and northern Australia has survived ... [It consists] of an intimate knowledge of plant and animal cycles, and contains details of the intricate connections between these, which the Aboriginal people have observed over thousands of years and passed down from generation to generation ... They reflect the deep Aboriginal philosophy that 'all things are connected', and that subtle natural linkages are present which can reveal much about climate and weather.'

An example given, from the southwest Simpson Desert, is the appearance of plovers before the onset of rain. (Elsewhere people rely on hearing the furtive honking sounds that plovers make before rain, I'm told.) When cockatoos are flocking or when a bearded dragon stands with its head erect are some of the other traditional signs of coming rain.

Cockatoos know!

In many places in rural Australia, if a flock of, say, ten squarking black cockatoos flies over, people will take note and count them. The forecast is made that there will be rain in ten days. Others might give the alternative interpretation, that ten days of rain are coming up!

And it all may well be true. When black cockatoos get restless and squark raucously as they fly off somewhere – it often does foretell the coming of rain. What they are doing is positioning themselves for a good feed. The nuts and seeds that they love to eat are easier to extract after rain has soaked them. The cockies are probably heading off to their feeding grounds in forests. A friend of mine regularly observes flocks feasting in his own eucalyptus forest during wet spells.

7th May 2007, 8am. The awe inspiring, highly unusual sight of an enormous flock of magnificent, yellow tailed black cockatoos. They slowly and majestically cruised over my house, like gently gliding stealth bombers, in group after group, all squarking with delight. I'd never seen such a big mob, several dozen strong and they were on a mission! Those black cockatoos must have been be singing for joy that the ten year drought was over!

Sure enough, not long afterwards my local district of Castlemaine scored more rain than anywhere else in central Victoria, with over 66 millimetres on just one night – 18th May. Rain fell steadily throughout the night and by morning winter creeks, bone dry for several years, were flowing again and Campbell's Creek was spilling over. My dam was overflowing. The total May rainfall of 152 millimetres proved to a record, the highest in 42 years. Those cockatoos would be really feasting!

Indian folk wisdom

Across rural India age-old folk wisdom is a continuing basis for making weather predictions and it was featured in a leading article for India's *Tribune* in 2003. Abhay Desai wrote that:

'The indicators would baffle scientific reasoning. These could range from fireflies appearing in the night forest, sparrows bathing in the dust and ants carrying pupae, to goats creating a din and refusing food, to the flowering of Neem and Babul trees—all suggesting impending rains.

Thus, long before the Indian Meteorological Department had predicted drought in Rajasthan, the Bhil tribes of the Thar were already prepared. They noted tell tale signs – the khair trees (Acacias) had become extra bushy and wild cucumbers sprouted everywhere. Other drought indicators elsewhere include crows cawing during the night, foxes appearing during the daytime and snakes climbing up trees.

Prolonged wet spells, on the other hand, are forecast by chameleons climbing trees, crows scratching their nests, peacocks wailing, lapwings laying eggs during the night, camels that keep facing northeast and frogs that start croaking before nightfall, amongst many other traditional indicators.

A grassroots-level body, the Ancient Rain-Prediction Network (ARPN), is conducting research. ARPN member P.R. Kanani, of Gujarat Agricultural University says that:

'The Indian Meteorological Department's predictions for specific regions are inadequate as they hold good for no more than three days and do not help in planning sowing operations.'

Astronomical prediction

The incidence of rain around the time of the full moon is commonly known, while a ring around the moon (being a ring of moisture) is a good sign that rain will come within the week.

The ancient Aztecs used their astute astronomical observations to help predict the weather, as people do in other places such as India, where farmers have traditionally used *panchangs*/almanacs for weather prediction. Panchangs, in use since the 4th century BCE, are based on folklore, astrology and ancient rituals. They are relied upon much more than government meteorological predictions.

In New Zealand author Ken Ring uses his knowledge of moon cycles to accurately predict weather and for this he is well known. In a radio interview (2UE Sydney, 6th Jan 2007) Ken was asked about climate change, an idea he rebuked, saying:

'The word climate contains in it the idea of cycles, and it is the cycles fed by the sun and moon that run the weather. For example the last time the drought was as bad was 1913, which was an exact moon-cycle equivalent to 2006. An equivalent match for 1913 was also 1983 which was also a bad drought year for Australia.

'The Moon creates weather because of its gravitational pull on the air. The theory is that the air has a tide, just as the sea does and just as a sea tide is measurable and predictable way in advance, then so should the air-tide be and therefore so too the weather.'

Ring says on his website that the meteorologists refuse to consider that the moon has any influence on weather. He says of his technique:

'We look at moon cycles and match present weather to historic weather that happened on same moon cycle times in the past .... In the past, ancient villagers kept their own records by planting sticks and marker stones around stone circles.'

Prediction by dowsing

Dowsing can be a useful tool for weather prediction. It can be a dowsing challenge that's easy to practice and develop confidence with. Results can later be confirmed by reading a rain gauge, or checking on a Meteorological website. (Rainfall is very variable and I would think it best to have your own rain gauge to try this with.)

American dowser Victor L Leroux has used his dowsing method to accurately predict weather. He wrote that:

'For years I have successfully used the pendulum to select the best time for my holidays, or for my activities of the day. I ask what the weather will be at this time, or that time, at this or that place, or at what time should I be doing this or that, if I wish to dodge the showers, etc'.

References

Dalton, Simone, 'Box Tree Buds Herald a Wet 2007' *The Weekly Times*, 29th March, 2006.
Weather bloggers at:
www.abc.net.au/science/k2/stn/march2000/posts/topic44431.shtm)
'Climate, Weather and Aboriginal Culture – A Precious Heritage'
www.bom.gov.au/iwk/climate_culture/Indig_seasons.shtml
Desai, Abhay, 'Quaint ways to predict rain' *Tribune*, 6th July 2003,
– www.tribuneindia.com/2003/20030706/spectrum/main4.htm
Ken Ring's website – www.predictweather.com
Leroux, Victor L, 'Weather Prediction by Dowsing', *The American Dowser*, via Dowsers Society of New South Wales newsletter February 1992.

3.3. Rain-making traditions

Universal rituals

Rituals for invoking rain have been around since time immemorial. They must have enjoyed some success to have continued for so long. From the rain dances still performed by the Hopi in southwest USA, to the beating of ceremonial rain drums in Ugandan rites, people have often used similar methods.

Rituals were often conducted on mountain tops. In prehistoric Mexico the rain god Tlaloc was worshipped on the summit of Mount Tlaloc, to the east of Mexico City. In native American tradition rain-making rituals were held at landscape power centres such as Rain Rock and Thunder Rock. During droughts Apache people appealed to the 'Controller of Water', who sits at the water gate and stops the water. Gary Varner quotes the Apache who say of the Controller:
   'He holds the rain. He lets it loose or shuts it off. You sing to him if you know his song. If people believe and learn his song, they can get rain.'

Ancient peoples created images to represent their water gods and goddesses and to invoke rain. Times of climate change were thus documented in Europe. A prolonged drought around 8000 years ago in south-eastern Europe saw a plethora of various artforms depicting the Mistress of Waters as the Water Snake, Water Bird and Bear Goddess, particularly in the drier parts of the Balkan Peninsula.

The goddess's all-seeing eyes stare out at us from the Neolithic ceramic vessels, in association with rain torrents – lines running parallel; sine waves that are either wavy or zigzag, or dotted bands; chevrons and meander patterns. The sacred artforms were also covered with snakes and we find serpents that are horned, or striped, in formations, or coiling around pots. Then there are the clay statues of naked rain goddesses holding bowls (the earliest Cauldrons of Rebirth, perhaps), all attesting to an obsession with water. Author Professor Marija Gimbutas commented that:
   'The desiccation of the climate during the sixth millenium BC, shown by palaeobotanical and geological research (and clearly revealed by the author's own recent excavation at Anza, Macedonia) is also reflected in

symbolic communication. The centuries long lack of water resulted in the creation of symbolic images related to streams and mythical creatures considered to be the source of water.'

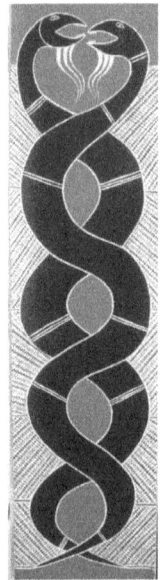

Serpent deities

The veneration of primordial water serpent spirits is widespread in the world and lies at the root of many spiritual traditions, including rain-making. Some surviving artistic images of sacred serpents have great antiquity. Sculptural forms in Central America depict serpents in clay, wood and stone; serpent mounds are found from Africa to Canada; and in Australia there are serpentine stone arrangements in Victoria, as well as a plethora of modern snake spirit paintings by Aboriginal artists.

Photo: Water Pythons, painting by Adam Henwood, from the Dreaming of Rex Wilfred, NT.

Chinese water dragons

In China, dragons have long been associated with water in the landscape, as well as weather, tides and water levels. Although they made fearsome adversaries, dragons were generally regarded as auspicious, the protectors and guardians of treasures, waterways, clouds, winds, and Heaven itself. One of four animals representing the cardinal points, the dragon is associated with the east, springtime and fertility. In Taoism dragons are benelovent, happy spirits. But after the arrival of Buddhism they were more often cast as snake-like and menacing.

In popular folklore the Long Wang are dragon kings, of great wisdom, strength and benevolence, responsible for rain and water bodies. Offerings to appease the Long Wang were made during droughts, as well as in times of serious storms, fog and earthquakes.

On the local level, water dragons are considered to govern the fertility of the fields and the prosperity of the people. But if a water dragon has been interfered with, havoc can be unleashed. Storms are said to be battling dragons, droughts are sleepy or angry ones and floods are the wrath of very cranky dragons. Dragons spend their winters underground and when they traditionally rise up to the surface in spring – on the second day of the second lunar month, coinciding with spring rain and the crop planting season – the occasion is celebrated with a day of welcoming processions and colourful dragon dances and festivities. A Western observer (de Visser) wrote in 1913:

'Dwelling in "dragon pools", when they woke up in spring and were seen coiling and revolving it was regarded as a creative force.'

To bring rain, dragons are said to fly up to the clouds and play with flaming pearls or balls, symbolising thunder. Dragons were regularly invoked in times of drought and the dragon's image would be taken out of the temple and paraded around to show it the damage done and to encourage precipitation. The dragon dance that's still performed at Chinese New Year festivities was originally a ritual of rain-making.

The forms that dragons take in Chinese art show conglomerate serpentine creatures that often have the horns of a deer, the head of a camel, eagle claws or tigers feet, fish scales and oxen ears. Like water, dragons are said to be coloured blue-green, have the ability to shrink and become invisible, and also be able to fill up 'all the space between Heaven and Earth.'

The early Chinese gods of the skies were also invoked by the people during droughts. The Master of Rain has a one-legged bird that can drink the seas dry. The Goddess of Lightning flashes her two mirrors to cause lightning, while the God of Thunder makes thunderclaps with his hammer and drums. In the rain-making rites of the Bronze Age Chinese (as well as in Indochina and Indonesia) ceremonial kettle gongs were once rung to emulate thunder.

Rain-making in Africa

In South Africa serpent spirits have traditionally been held responsible for rainfall. A legendary giant antelope-headed snake, *Inkanyamba*, lives on the summit of Mount Mpendle and brings rain by 'traveling from mountain to mountain in order to copulate in water bodies'; while near

Genaadeberg, a trickster deity *Kaggen* is 'known to have danced with a rain snake' in a rain making ritual, says Le Quellec.

Rain-maker experts of the southern African Bushmen go down to a water hole or springs said to be inhabited by a *rain bull* (water spirit). This they somehow 'capture' and lead around the countryside to another site where 'it is cut/bled' (in other words the rain is symbolically made to fall), writes author Paul Devereux.

Rain-making in India

In Indian and Buddhist tradition kingship and rainfall were divinely linked. One of a king's roles was to invoke rain when needed. A righteous king would ensure rain in due season and if rains failed it was the king who got the blame. In order to maintain his position and power a king had to convince his subjects that he had the ability to produce rain.

In Sri Lanka the mountain at Mihintale and the rock at Sigiriya were two of the main centres of ritual worship where pre-Buddhist Sinhalese kings held festivals involving both rain-making and fertility, says Disanayaka. (The later Buddhist kings shifted their ritual focus from mountain tops to Bodhi trees.)

Rain-making in Europe

There are delightful descriptions of rain-making ceremonies from the British Isles, where the playful frolic of ordinary folk was the order of the day. At a rain-making rite at Gellionen Well, near Pontardawe, West Glamorgan, people danced on the green throwing flowers and herbs at each other, singing jolly ballads and playing 'kiss in the ring'. The group leader would then go to the well and cry out three times 'Bring us rain!!' Everyone then filled up bowls with well water and flung it around or took it home to sprinkle on the garden. Apparently it never failed to rain after such an event.

In Lincolnshire tradition, the Boundels Stone could make it rain when it was hit with hazel rods. Another stone close by, when beaten, could make corn grow. The locals held an annual feast around the pair, which they whipped 'till everybody went wicked wi' prosperity,' quotes Bord.

A French healing well at Folle-Pensee, Morbihan, was once used for weather rites. A libation of its water used to be poured onto a rock called Merlin's Step to produce a storm; this was last recorded as successful during a drought in the 1930s.

Rain-making in Australia

Aboriginal water specialists preside over rain-making rituals. Archeologist Aldo Massolo describes the practice in south-eastern Australia, with a ceremony involving imitative mimicry – the sprinkling of water, furious shaking of branches and loud chanting suggestive of rain and wind. In the 1970s Massolo wrote that:

'Rain-makers were probably keen students of nature and must have known exactly when to start their practices, since they were often seemingly successful in achieving their objectives.'

At Goanna Headland, near Evans Head, the Bundjalung people of northern New South Wales still go to visit a sacred cave in times of drought. At this rain *djurebil*, or increase site, they sing the required songs asking for rain. Floods down the Evans River have been attributed by them to a violation of the sacred cave.

I also heard recently of a group of Bundjalung women around Mount Warning, north of Evans Head, who were heading off to a Weeping Site to sing up some rain, or cry for it, perhaps. The original name for the mountain, *Wollumbin*, the 'cloud catcher', is an apt title. The 25-million-year-old, rainforest-clad volcano is the wettest place in the state.

In much of Australia permanent water holes, the homes of serpent spirits, are where rain-making rites have been held, sometimes with several tribes participating. Particularly important in arid areas, such inter-tribal rain-making ceremonies were important social occasions too.

At Mutawintji National Park, in western New South Wales, large gatherings of Malyankapa, Pandijkali and Wimpatja people once took place. Petroglyphs there have been dated to over 8000 years of age. Initiation rites were held at Snake Cave and other places were used for rain-making ceremonies. It is still a tribal meeting place, but too arid now for habitation. In north-west Northern Territory the Walpiri tribe have for many generations gathered at Mawuritji, near Lake White, together with the

Kokatja, of Western Australia, and the Ngardi people for initiation and rain-making ceremonies, Tindale reported.

In the south Kimberleys rain-making ceremonies are typically performed at *jila* – sacred water holes that are home to *pulany* / snake spirits. Local pulany have the power to generate rain at any time and do so if they are disturbed. It is to the jila that men go to control rainfall, by persuading the pulany to 'get up'. It is from the jila that the knowledge of songs and dances relating to rain-making arise and are given to people in dreams. To induce rain, special songs are sung and body designs are worn as people carefully dig around in the jila, throwing mud up to attract rain.

A senior rain-maker, a *yiliwirri*, uses his power to placate the pulany or make them 'get up'. This is demonstrated in his ability to make winter rain, whip up thunderstorms and lightning, and divert storms or cyclones away from or towards particular areas. This can be also done in other areas where pulany reside, such as the sea, says Peter Tapirri.

The Kimberley region is also home to Wandjina spirits. Often depicted with striking rayed head halos in rock art, the Wandjina co-exist with *Unguds* / rainbow serpents in their sacred water holes and are also associated with Earthly fertility. Wandjina are held directly responsible for rain, whereas the serpent spirits are more involved with groundwater. The Wandjina's head rays represent lightning, their voice is the thunder. Their mouths are not shown, because if they were to open a flood would ensue! Paintings of Wandjinas and Unguds used to be restored annually at the start of the wet season, to ensure the return of life-bringing rains.

When rain is needed, the Wardaman of northern Australia say that 'Rainbow has to occasionally give Lightning Man a nudge to make it rain'. Monsoon rains usually follow the annual October ceremonies that are held there at the end of the dry season, explains Bill Yidumduma Harney.

The Nyigina way
Paddy Roe describes indigenous traditions of the southern Kimberleys in his book *Reading the Country*. Roe was born in 1926, near the seven or so Mimiyagaman springs area of Roebuck Plains. Born on a nearby hilltop, the springs are his *jila* and his sacred duty is to be their guardian. Roe, an Nyigina tribal elder, knows a thing or two about rain-making (and is also sought after by white people as a water diviner). He tells a story of

going to one of the smaller of the Mimiyagaman springs with his uncle for a rain-making ceremony. At the Nilababa ('ear hole') spring a central tuft of grass growing out of the water is said to be the ear hairs of the resident rainbow serpent. Here the two men first removed their clothes, then waded out into the water asking for rain to fall on a particular sheep station. The appropriate rain song was sung. Then a spear was plunged into the centre of the pond, deep into the white clay bottom. It was then retrieved and the white clay carefully cleaned off.

Not long afterwards, a few white clouds (which the white clay had no doubt represented) gathered over in one direction, where it then rained. On checking the next day the men found that the dry paddock in mind had indeed received the requested rain, but beyond the fence line it was still dry. It had rained exactly on the area specified and nowhere else.

Paddy warns of one large spring – 'The Boss' – that is the home of the *yungurugu*. In it a large grassy patch is said to be the goatee beard of the yungurugu and if anyone damages this hair there will be dire consequences, such as dangerous electrical storms or cyclones. Nobody touches 'The Boss'. Only a *maban* (- a 'doctor') would risk, or indeed be able to see or communicate with the yungurugu. A Maban could also command the snake spirits to avenge the violation of Aboriginal laws, for yungurugu are able to leave their springs and rise up into the air and follow people. Dreaming stories about this happening help to reinforce the imperative of keeping to the law.

Rain-making art

Rain Dreaming sites in the Northern Territory often have ancient rock engravings that feature abraded grooves. When rain ceremonies are conducted the grooves are deepened and such designs may also be painted in ochre on the bodies of performers.

Rain Dreaming is also depicted in the paintings of modern Aboriginal artists. For instance Western Australian artist Johnny Mosquito is the custodian of the site of Puntukutjara, south-west of Balgo, and is the senior rain-maker in the Balgo community. This theme is evident in many of his paintings, which often depict Water Dreaming stories. To this day John often visits the region during the dry season to perform rain-making ceremonies. Dinny Tjampitjinpa Nolan, who was born about 1920 near present-day Yuendumu in central Australia, is another such artist. As a highly respected senior custodian for the Walpiri people, he is officially recognised as a leader of rain-making and Water Dreaming ceremonies. He oversees corroborees, paints the bodies of those taking part in the ceremonies and is renowned for his powerful singing voice.

Rain songs
Aboriginal rain songs have been performed during Rain Dreaming ceremonies since time immemorial. They are owned by particular people and may be linked together as song lines or Dreaming tracks – that is, groups of sacred sites connected together by both songs and pathways. Researcher Jane Anderson explains that:

'The Kaytetye Akwelye Rain Song series is the one most identified with the powerful rain-making sites in the Arnerre land estate, to the west of Neutral Junction Station in the Northern Territory. This song series can be traced along a track going from west to east, travelling close to the Kaytetye Tara camp area.

'These songs were (originally) given in a dream by an ancestral rain figure, through a mediating spirit, to a corporeal traditional owner of Arnerre country. ... [But] ceremonial life, particularly initiations at the camp at Tara, had been suspended for six years during the presence of an Arrernte pastor, but after his departure in the mid-1970s, the ceremonies had been reinstated.'

When first recorded by white people in the late 1970s the songs were restricted to female ears only. Anderson said that:

'The five-hour performance was sung away from view of the rest of the community in a tin shed which had been used for general meetings and as a church. Some women stood outside the shed to discourage visitors.'

But Anderson was later surprised when, in the mid-1980s, one of the song's owners sang them on the radio for all to hear. These days a new generation of Aboriginal people are often more keen to share their sacred traditions in a re-blossoming of cultural pride.

Rain making in New Zealand

Rain, the 'tears of Ranginui', was always considered to be blessed with goodness and compassion by the Maori, who were great agriculturalists. In their gardens they placed stone and wooden statues to represent the *atua* (gods) and act as the *mauri*, or resting place of the life principle of the garden. Here, at crop planting time, the *tohunga* /priest recited ritual incantations, and again at harvest time.

If rain was needed, the tohunga would swing a wooden bullroarer to ritually bring rain to the crops. And across the Tasman, writes Mircea Eliade, some Aboriginal traditions regard the sound of sacred ceremonial bullroarers as the embodiment of lightning. The Hopi Indians in America also swing a bullroarer during ceremonies, the whirring sound being said to attract wind and rain.

References

Gimbutas, Marija, *The Goddesses and Gods of Old Europe – Myth and Cult Images 6500-3500 BC*, University of California Press, USA, 1982.
Land of the Dragon – Chinese Myth, Duncan & Baird, UK, 1999.
Encyclopædia Britannica 2004.
The World of Myths, Vol 2, British Museum Press, UK, 2004.
Le Quellec, Jean-Loic, *Rock Art in Africa: Mythology and Legend*, 2004, Editions Flammarion, Paris, at: www.sommerland.org
Devereux, Paul, *The Sacred Place*, Cassell & Co, UK, 2000.
Disanayaka, J B, *The King as Rain-Maker*, Ministry of Environment & Parliamentary Affairs, Sri Lanka, 1992.
Varner, Gary, *The Lore of Sacred Water*- www.druidnetwork.org/articles/garyvarner.html#f1
Bord, Janet and Colin, *Earth Rites*, Granada, UK, 1982.
Bord, Janet and Colin, *The Secret Country*, Paladin, UK, 1978.
Mootwingee – NSW Department of Education and Training Sites and Scenes 1999, on-line.
Tindale, N B, *Aboriginal Tribes of Australia*, 1974.
Tapirri, Peter Clancy, 1999 interview, in *'Living Water'*, Centre for Anthropological Research, University of Western Australia (on-line)
Cairns, Hugh and Yidumduma Harney, Bill, *Dark Sparklers,* published by H Cairns, 2003 – PO Box 83 Merimbula NSW 2548.
Roe, Paddy; Benterrak, Krim and Muecke, Stephen; *Reading the Country*, Fremantle Arts Centre Press, Western Australia, 1984.
Aboriginal art:
www.aboriginal-desert-art.com.au/artists/dinny_nolan_tjampitjin-pa.html

Anderson, Jane, '*The politics of context: issues for the law, researchers and the creation of databases*' PARADISEC (Pacific And Regional Archive for Digital Sources in Endangered Cultures) AIATSIS at: www.conferences.arts.usyd.edu.au/viewpaper.php?id=61&cf=2

Eliade, Mircea, *Australian Religions, an introduction*, Cornell University Press, USA, 1973.

Nayutah, Jo and Finlay, Gail, *Our Land, Our Spirit – Aboriginal Sites of North Coast New South Wales*, North Coast Institute for Aboriginal Community Education, 1988.

3.4 Sky water harvesting

Outer space origins

There is evidence that much of Earth's water originated from outer space in the form of icy comets, slightly damp meteors and other objects that bombard the atmosphere. As these have also been known to contain amino acids, the building blocks of life, this adds to the notion that life on Earth was originally seeded from beyond and in association with water. The Earth was heavily bombarded by meteoric material from 4.5 billion to 3.8 billion years ago.

As for how old our water is, the dating of meteorites has given an age of 4.6 million years for our solar system. Earth seems to be about the same age. Earth's oldest sedimentary rocks, having been formed in a watery environment, are only around 3.9 million years old, but there are zircon crystals in Western Australia that were created in the presence of water and have been dated back to 4.4 billion years.

Rivers in the sky

In 1993 researchers announced a new discovery: of a network of rivers of atmospheric water vapour, some carrying as much water as the Amazon River, flowing as high as six miles (nearly 10 kilometres) up. For years climatologists have mapped the flow of water vapour in the atmosphere using weather balloons. They have observed huge sheets of water vapour evaporating in the steamy equatorial regions. These drift off towards the Poles, sometimes wandering east or west as winds and air pressure varies.

Reginald Newell, an American climatologist, was surprised to discover long, narrow filaments that look like rivers meandering and flowing into one another. Atmospheric rivers have since been found all over the globe, the longest being 4,000 miles (6440 kilometres). Newell reported that the sky rivers:
'... are generally about 150 miles [240 kilometres] wide and just a mile [1.6 kilometres] deep, with 350 million pounds [about 16,000

tonnes] *of vapour flowing past a given spot each second. The rivers are transient, but at any given time there are at least a few of them in the atmosphere, typically around five in each hemisphere. In general, the vapor rivers head for the Poles, but on the way they can get deflected toward the east by Earth's rotation and flow, for instance, from Africa to Australia.'*

Water vapour and condensation

Water vapour is ever present in the atmosphere. The hotter the air, the more water it can contain. Water vapour naturally condenses on cold surfaces into liquid water and in nature we call it dew. When clouds reach up high into cooler zones, rain is precipitated.

For condensation to occur on a surface it must be cooler than the water vapour. The water molecule brings a little heat with it, so when it condenses onto a surface, a net warming occurs on that surface and, as a result, the temperature of the atmosphere drops very slightly. The dew point of an air parcel is the temperature to which it must cool before condensation begins.

This simple process of condensation can be used to generate water for human use. There are vast amounts of water vapour available for harvesting in the Earth's atmosphere. It is constantly evaporating from oceans and rising up to about one hundred kilometres above the surface. Even in places that don't seem humid, water vapour will be present. In fact it fills an average of 60 percent of the Earth's atmosphere, totalling some 8000 cubic kilometres of water vapour.

Ancient structures made for the collection of sky water by condensation, including air-wells, fog-fences and dew-ponds, have been discovered in various parts of the world. Various modern devices have also been available commercially for a few years now, while other prototypes are (hopefully!) being fast-tracked.

Air-Wells

People have been extracting water from air for millennia. This was shown in 1900–1903 by archeological discoveries at Theodosia, a Byzantine city that flourished about 500 BCE, in the driest part of the Crimea. Here numerous clay pipes, about 10 centimetres (3 inches) in diameter, were found leading to wells and fountains in the city. The pipes were traced back to a nearby hilltop, where 13 huge heaps of loosely piled limestone, each about 12 metres tall and 30 metres, square were located. This system of air-wells has been estimated to have produced as much as 63,000 litres of water daily, a *Popular Science* article of 1933 reported.

People in the 20th century experimented with air-well technology, using giant domes or piles of rocks to capture moisture from the air. Belgian inventor Achille Knapen at Trans-en-Provence constructed an air-well that looked like a gigantic perforated beehive 14 metres tall. It allowed warm air to enter and dump its water load on contact with the chilly concrete interior, which was studded with rows of slates. Despite higher expectations it produced only 23 litres of water per night, however. The design had been inspired by bio-climatologist Leon Chaptal, who built his own small air-well near Montpellier in 1929. Nelson writes that:

'Chaptal found that the condensing surface must be rough, and the surface tension sufficiently low that the condensed water can drip. The incoming air must (ideally) be moist and damp.

'[Chaptal's] pyramidal concrete structure was 10 x 10 x 8 feet [three metres square and 2.5 metres] in height, with rings of small vent holes at the top and bottom. Its was filled with pieces of limestone [5-10 cm] that condensed the atmospheric vapour and collected it in a reservoir. [The yield varied with the seasons and] ... its maximum yield was 2.5 gallons [5.5 litres] per day'.

Dew-ponds

Shallow, never failing dew-ponds in Britain have also been around since ancient times and a few survive on the ridges of England's Sussex Downs and the Marlborough and Wiltshire Hills. Dew-ponds are not connected to groundwater, nor are they designed to catch rainfall, yet they always contain some water, thanks to condensation from night air. Wherever sea mists roll in, or fog hangs around a valley, is an ideal locations to site a dew-pond.

Dew-ponds typically measured 9-21 metres (30-70 feet) across, with a depth not exceeding 1-1.2 metres (3-4 feet), an article in *Scientific American* (May 1934) explained. Writing on the subject around 1900, C. J. Cornish said that:

The Wisdom of Water

'Dew-ponds are mainly found on porous soils where the rainfall is about 40 inches [one metre] per year, and the annual evaporation from a body of open water is in the range of 18 inches [450 millimetres].

'The shepherds say that it is always well to have one or two trees hanging over the pond, for that these distil the water from the fog. This is certainly the case. The drops may be heard raining on to the surface in heavy mists.'

In his book *Neolithic Dew-Ponds and Cattleways* (1907), Arthur J. Hubbard described the process of making a dew-pond thus:

'The gang of dew-pond makers commence operations by hollowing out the earth for a space far in excess of the apparent requirements of the proposed pond. They then thickly cover the whole of the hollow with a coating of dry straw. The straw in turn is covered by a layer of well-chosen, finely puddled clay, and the upper surface of the clay is then closely strewn with stones. Care has to be taken that the margin of the straw is effectively protected by clay.

'The pond will eventually become filled with water, the more rapidly the larger it is, even though no rain may fall. If such a structure is situated on the summit of a down, during the warmth of a summer day the earth will have stored a considerable amount of heat, while the pond, protected from this heat by the non-conductivity of the straw, is at the same time chilled by the process of evaporation from the puddled clay. The consequence is that during the night the warm air is condensed on the surface of the cold clay.

'As the condensation during the night is in excess of the evaporation during the day, the pond becomes, night by night, gradually filled. Theoretically, we may observe that during the day, the air being comparatively charged with moisture, evaporation is necessarily less than the precipitation during the night. In practice it is found that the pond will constantly yield a supply of the purest water'.

'The dew-pond will cease to attract the dew if the layer of straw should get wet, as it then becomes of the same temperature as the surrounding earth, and ceases to be a non-conductor of heat. This practically always occurs if a spring is allowed to flow into the pond, or if the layer of clay (technically called the 'crust') is pierced.'

More modern concrete structures to collect water vapour, derivative of the dew-ponds, have also achieved some degree of success. One imagines that, in places like Australia, where evaporation rates are much higher than in Britain, the addition of a shadecloth cover over the condenser structure, weighted down in the middle, would be useful.

Fog-fences

Along mountain ranges close to coastal regions, fog is sometimes the only source of water. Water can be extracted from the fog by intercepting it with special huge fences. This has been done along the Pacific coast of South America, where locals call it 'harvesting the clouds'. The ideal locations for these 'fog-traps' are described by Nelson as:

'arid or semi-arid coastal regions with cold offshore currents and a mountain range within 15 miles [about 24 kilometres] of the coast, rising 1,500 to 3,000 feet [50-100 metres] above sea level. ... [For example] a group of 50 fog-traps made of plastic mesh stand atop a 2,600 foot [800 metre] mountain [at Chungungo, Chile] and they collect up to 2,000 gallons [9000 litres] daily.

'... Mesh occupying 70 percent of the space is most effective for trapping fog droplets. Two layers of mesh, erected so as to rub together, optimise the collection of water in PVC pipes attached to the bottom of the nets. ... Chungongo now has its water needs met and there's even enough water to begin to reafforest the area, which will then become totally self-sustaining. The previous forests there were nature's original fog harvesters.'

Let the people drink thin air!

Letter from Maureen Brannan:

'I heard on Radio National News last week that south-east Queensland Mayor Ron Clarke was sending a delegation to Russia to investigate this air-to-water technology, which the Russians have embraced, with machines apparently producing 200 megalitres a day. (Machines are also being mass produced in Korea and elsewhere.)

'In this news item, he mentioned Peter Beattie's comment the week before, that his government HAD to build the Mary Dam because of South-East Queensland's expanding population, and because, quote, '... people can't drink thin air ...', and Ron Clarke made the comment that people CAN drink thin air – through this simple technology that has evolved from humidifiers.'

Water from wind

Water can be harvested from wind with a unique windmill invented by Max Whisson, a retired West Australian doctor. The unique 'Whisson windmill' has sets of multiple aerodynamic blades arranged vertically, which can whirr into action at the merest puff of wind. These harness the energy of the wind to power a refrigeration unit that cools and condenses atmospheric water vapour. It's all very simple.

Still in the prototype stage, no large-scale investment has been forthcoming, so prototypes have been put together with scrap metal in a shed in Subiaco, Perth. Whisson enjoyed some television exposure in May 2007 and said that the windmill is now undergoing further development at the University of Western Australia. The commercial 'Max mill' is expected to be sold as a four square metre device whose output could vary from 4000 to perhaps 8000 litres of fresh water daily, Whisson said. With a price tag of about $40,000 the 'Max mill' will not be cheap, although, if claims are correct, one unit would be enough to provide sufficient water for perhaps ten households. So a shared investment in a 'Max Mill' by clustered housing groups would, theoretically, be extremely cost-effective.

The spray turbine

In 2002 one of Britain's leading inventors was given a government grant to develop the world's first rain-maker machine. Professor Stephen Salter, the Edinburgh University engineer renowned for 'Salter's ducks', went on to develop what is known as his 'giant egg beater' machine. *The Times* reported that:

[the] ... *'rain-maker uses wind power to drive a 200 feet [60 metres] high turbine that sucks water out of the sea, and turns it into water vapour and sprays it out into the atmosphere, creating clouds ... The machine uses a Darius turbine, a vertical axis turbine that spins around, that's driven by the wind. The turbine blades have water pipes inside them, with an inlet just below the surface of the sea. The centrifugal force of the spinning blades sucks the water out of the sea and propels it nearly 200 feet [60 metres] up inside the blades. It is then forced out of nozzles, creating a spray that turns to vapour ... The best times to use them are when some clouds are already present.'*

The resulting rain could transform coastal desert areas and well as potentially reduce conflict in places like Palestine and Israel, Salter enthuses. Let's hope that the 'eggbeaters' don't end up as did the 'Ducks', which generated huge public interest in the 1970s when first developed. But the ducks became dead ducks. Killed off when the British Government decided to pursue nuclear power instead.

Modern rain-making

Techniques of aerial 'cloud seeding', whereby silver iodide, dry ice or ordinary sea salt is dropped into clouds from a plane, have been trialled in Australia and elsewhere for over 60 years. However rainfall sometimes actually decreases after cloud seeding and gains in some areas have caused a rain shadow in others. Accuracy for a particular target area is also pretty much impossible. By the early 1970s most Australian trials had been abandoned.

A recent conference held in Melbourne has just re-focused on the subject. A spokesman from Nevada's Desert Research Institute told attendees that cloud seeding is commonly used in the western United States to increase rainfall for power generation and farming. Participants were enthused enough to form a national taskforce to further investigate cloud seeding and a trial will soon begin in south-east Queensland.

An important message brought to the conference from an Australian expert was that cloud seeding is a waste of time during a drought. In times of normal rain, cloud seeding can enhance rainfall. There has to be some cloud already present before it can be successful. (And it is often likewise with ancient psycho-spiritual techniques.)

More esoteric rain-making methods have been developed in the meantime, often inspired by the insights of people like Wilhelm Reich and Rudolph Steiner. It was Reich who made the observation that times of drought were always accompanied by sluggish atmospheric conditions, the atmospheric ether (which he called 'orgone' energy) becoming congested with 'deadly orgone radiation' (DOR, or 'bad energy'). Reich conducted a great deal of research in the early 1950s on how to clear toxic atmospheric conditions. He lived downwind from nuclear testing sites, so there was plenty of DOR around. (Movie star John Wayne and cowboy-and-western movie film crews were also downwind from the testing site and many died of cancer as a result, Wayne included.)

Reich's prototype 'cloud-buster' device was used to draw atmospheric energy around the sky and to make it rain. It worked, but it needed a good driver! Nowadays, 'energy farmers' also use techniques such as biodynamics, 'Towers of Power' and the practice of Agnihotra, to energetically harmonise the general environment and, as a side-effect, make it rain more. And, continuing ancient global traditions, ritual rain invocations are also still practised by people today.

Energy balancing and rainfall
by Tim Strachan

'It seems to me that 'rain-making' or 'cloud-busting' are not quite right in describing the process of encouraging rain. The right amount of rain will appear when there is an energetic balance in an unpolluted environment, as night follows day. Just as with any disease, the problem is a lack of flow or communication due to blockages, which can occur on many levels: pollution, electromagnetic congestion, abuse of the aquifers and water table, and human energetic interference and distress, amongst other factors.

Successful rain-making really occurs when the environmental energies are balanced and this is what Wilhelm Reich did repeatedly in various 'cloud-buster' trials in the early part of the twentieth century. He devised ways of balancing energies by building up orgone in the atmosphere, and by draining deadly orgone with a pipe connecting his device to running water, the only element that can properly deal with such energies.

Reich's "cloud-busting" work was furthered by people such as James Constable with his etheric rain engineering, and, more recently, by dowser Don Croft with his simplified portable energy balancers. Like Reich's, these have several copper pipes about 1.8 metres (six feet) long embedded in an orgone matrix, which is a mix of organic and inorganic material, in the form of resin and metal shavings. It's an improvement on Reich's device as it appears to absorb the deadly orgone from the atmosphere and re-broadcast it as a benign energy. The photo shows one of these standing on a highly paramagnetic material (special rock dust) in a copper spiral (industrial heater element) in our garden. These are additions to Don's device that seem to improve its functioning.'

'Cloud-busting' results

'Up to mid-2004, we lived in Bucketty, an hour north of Hornsby (Sydney), a normally dry area which is ecologically fairly undisturbed. We had the 'cloud-buster' there from the middle of 2002; that is, through some of the worst times of drought. Our property was designed according to geomantic principles with a standing stones, stone circle, ponds, flow-forms, solar electricity and heating, etc, all of which helps in modifying energies for the better. We had brought back the frogs and all kinds of animals and even locals felt the difference when they visited us.

In spite of the drought we had more rain there those 24 months than was normal. Our tanks kept full and rain patterns changed. Originally the rain came up from Sydney and either stopped about 10 kilometres away or just curved away from us. But we started to get that rain and another pattern also started up, with rain to the west of Sydney moving up to our area, often bypassing Sydney. And what also happened, somewhat unbelievably, is that rain clouds would form above our area, rain would fall, but you could see that it was like a pancake over us, right around the horizon. The sky could be blue all around the pancake of cloud

I believe it made quite a difference in this area, and the effect was cumulative. This region is responsive to small energy differences because

> it has not been put deeply out of balance, but a city like Sydney, which is increasingly blocked on an energetic level, would need quite a number of these devices spread around to make a difference. It seems that if there is humidity or rain about, it is more likely to be drawn in by this device. Likewise, if there is too much rain, I believe that this will be reduced. In other words – metal shavings a harmonising effect occurs.
>
> Reflecting back in 2007, we left Bucketty in mid-2004 and left the cloud-buster there with the new owners. There was also another one which I helped a neighbour make on his property some 1.5 kilometres away. We were passionate about energies and balance and living sustainably, and our successors and neighbours perhaps less so. But it did seem that the positive effects noted above had reduced in the years since we left. I can only imagine that this is a mysterious feature of 'energy technologies', and I am reminded of the curious fact that over 100 years ago Keely devised an 'over-unity' motor which put out more energy than was put into it. But it only worked when he was present. There was no fraudulent practice going on, it simply was somehow connected 'etherically' to him and his field. In a similar way, I believe that the attention and state of the 'energy body' of whoever is working with these technologies is crucial to their proper operating. To me that makes sense, as we are not separate from our environment, and we affect everything we touch significantly, for good or for ill.'

Rain-making in biodynamic farming

Biodynamic farming was inspired by the clairvoyant insights of Rudolph Steiner. Its practice involves holistic farm management, working with the energies of the Earth and planetary influences, as well as special balancing sprays for soil and plants. Biodynamically produced food is high in nutrition, flavour and life-force.

Biodynamic researchers Hugh Lovel and Hugh Courtney have pioneered a gentle technique of atmospheric energy balancing and clearing, which involves spraying the 'BD preps' (special biodynamic preparations) in a particular sequence over several days. Four days after doing this, they say, it is bound to rain! The sequence is to spray the BD preps 500, then 501, then barrel compost*, 508, and 500 again. Whenever hazy dry conditions have set in, Lovel repeats this sequence, always with success, except once when Mercury was retrograde.

Courtney does it usually two or three days before full moon, when it tends to rain more, and ideally when the moon is in a water constellation. Lovel begins by spraying barrel compost onto soil at sunset and sunrise, accompanied by a light spray of the herb horsetail for balance. Lovel claims to have made it rain across a 32-160 kilometre (20-100 mile) radius area with this technique. These days he uses the easier method of broadcasting the BD preps remotely by radionics (- advanced dowsing techniques). Using a Malcolm Rae radionics machine he does morning and evening broadcasts of the same sequence, using an aerial photo as a witness and with homoeopathic potencies of the BD preps that the land requires.

---

* Barrel compost

This is a special BD compost, made from pure cow manure that has been composted in a barrel buried in the ground. The manure heap has six BD preps inserted into it. A recipe devised by Maria Thun is an improved version of this. It requires cow manure, ground egg shells (from healthy organically raised hens) and basalt dust. These are combined with the preps 502–507 and allowed to compost for 12 weeks. One barrelful is enough to spray onto 880 hectares (2000 acres). It makes a very good pioneering spray when converting to biodynamic methods.

Prayer and rain invocation

People who live in arid areas around the world, such as the Hopi Indians in America, have long used prayer in their invocations for rain. Modern mainstream religions follow the tradition. In Australia 'probably every drought since 1865 onwards has had a special day set aside for prayers', Keating writes, although not all the clergy have been happy about it.

Prayers have been offered from many different religious groups. For instance, a day of rain prayer was called by Church leaders for 1st December 2002. Then, on 19th January 2003, Islamic rain prayers were offered up in the Hunter Valley of New South Wales from the dry bed of the Goulburn River at Denman. But the outcomes of rain prayer are not usually very spectacular. However, events in 2007 may 'prove' its worth.

The Wisdom of Water

On 20th April 2007, Australia's prime minister John Howard asked the nation to pray to end the dire water crisis. Unless heavy rain fell in the next six to eight weeks, *The Age* reported, the hard-hit agricultural regions in the Murray-Darling Basin were facing zero water allocations for crop irrigation in the coming year (after 1st July). Howard encouraged all people to seek divine intervention to avert the imminent disaster.

A group of new-age entrepreneurs also took up the call and organised a synchronised rain visualisation and invocation event, focussing intentions 'centered around our intimate connection with nature'. They launched their own 'national rain day' on 8th May 2007, 11 a.m., on the lawns of Parliament House in Canberra and at the Botanical Gardens in every capital city. Afterwards a jubilant report was posted at their website, stating that:

'A week after this historic event many parts of the country received the blessing of abundant rain, which has filled dams and flushed rivers and creeks, nourished land and livestock, and the spirits of many people. This was not mere coincidence. It was a co-inside-dance of the hearts of all who imagined rain and directed love and gratitude that rain be received, who, quite literally, 'prayed rain'.

I hate to be a dampener, but while this sounds like a grand achievement, one must remember that the drought had already been broken by good rain in Victoria and western New South Wales well before the prayer call. Early to mid May stayed pretty dry but towards the end of May some good falls were widespread, however. The 9th May edition of *The Weekly Times* carried the story 'Late rain saves the month' and it was juxtaposed beside an image of the old *Herald Sun's* front page where Howard was calling us to 'pray for rain' in April. Even though he is probably not intimately connected to nature, Howard did suggest to the nation to ask for rain earlier than the new-agers. Or perhaps we should be thanking the Rabbinical Council of Victoria, who asked Jewish congregations to recite regular prayers for rain in the south-east of Australia from 12th May 2007 and ideally every week afterwards? Or was the National Council of Churches responsible, with their 26th November 2006 call for a national day of prayer for rain across the country? It was, after all, followed by good summer rainstorms in much of southern Australia.

Unless there is a proper study of rain events in relation to prayer events, we will never know if there is a correlation between the two, nor exactly how, perhaps, our prayers might best be directed. Certainly in the first half of 2007 many prayers went up and much rain fell down. By combining trials of cloud seeding and other more esoteric approaches it might be possible to make some impartial assessments and comparisons of modern rain-making methods.

Indigenous people, for instance, know not to ask for too much in their rain-making rites. In 2007 there might have been too much prayer. Much of Victoria and inland New South Wales enjoyed excellent autumn rains. By early June the Darling River had even begun to flow again! South-east Queensland also finally got a good drenching, its best in 18 months. On 8th June a massive storm front hit central coast New South Wales and lashed with ferocious intensity over several days. Mangrove Mountain had 300 millimetres on just one night and half it's annual rain in four days. The Hunter River flooded at its highest level since 1971. There was mass destruction and nine people died. Many areas were without power for days. While enough rain for another year's supply ended up falling in many reservoirs (some of which had been down to only 10 percent full), water supplies were cut or restricted to many areas, due to lack of power for pumping, a water shortage during an abundance! Twenty nine sewage plants also had no power for several days and had to release raw sewage into waterways.

When dowsers prayed for rain

Many dowsers have a well-developed trust and acceptance of the universe's ability to respond to requests. They may dowse to better understand a situation, then ask for problems to be resolved, in the interests of the highest good for all. This approach has been found to result in sometimes miraculous healings and sought-for events.

Those attending a West Coast USA dowsers' conference at Santa Cruz, California in July 1985, harnessed their group energy to avert what could have been a catastrophe. A huge bushfire had been lit by an arsonist on Sunday 7th and was raging near to the city, fanned by 48 km/h [30 mph] wind and 38 degree Celsius temperatures. The fire was predicted to burn wildly across to the coast for two weeks or more and no rain was expected to fall at this time of the year.

The dowsers decided to act. After dinner one hundred of them went to a redwood grove and held a special 15-minute rain service where they prayed for rain. Following this there was an immediate change in wind direction. That night an unexpected shower of rain fell in a nearby town. The next few days there was a reduction in smoke levels and a little good news filtered in to the intense conference activities. The fire was holding, they were relieved to hear.

By Wednesday morning people awoke to a thick, cool fog. At noon they held another, briefer, rain service, this time in the cafeteria, with 300 dowsers attending. Thursday was cool but sunny as people started their journeys home. Newspaper reports were then seen and noted. 'Drippy fog checks fire' read *The San Francisco Examiner* headlines. *The San Francisco Chronicle* reported that Wednesday's much lower temperature and humidity greatly helped contain the fire. The San Jose Mercury News reported that:

'a band of clouds blanketed Santa Cruz mountains on Wednesday, calming the Lexington fire and giving crews a chance to get a better grip on it ... The weather service doesn't know exactly why the clouds which formed over Arizona and Utah, moved into the area Tuesday night. That normally doesn't occur this time of year.'

Geomantic rain-making

Geomancers work with the subtle energies of the Earth and universe, in order to restore, balance or maintain Earth harmony between people and planet. They are the shamans, the spirit intermediaries and landscape protectors of their tribes. Geomancers are found in all cultures and corners of the world. The Australian Aboriginal law of the land lays out geomantic responsibilities for everyone, while the people of high degree have direct access to the spirit world. Modern geomancers continue these traditions and many dowsers take on the role too.

When rain-making rituals are performed at my Towers of Power (paramagnetic antennas) it usually rains soon afterwards. But I am not greedy, so I don't do them often. When the drought has been dire, I have gone to my Towers to ask for rain, making harmonic resonant sounds and singing special songs to the Spirit of Water. An offering of a little piece of gold doesn't go astray either! It's always best done when some cloud is apparent and I wait until I get a strong intuitive impulse to do it. Good results can be had when a group of people conduct a ceremony. The more people, the better. The power of several focussed minds and hearts is awesomely huge and potent!

Dan Winter, an American energy worker, suggests that if we want rain we should do some geomantic detailing work on our landscapes such as reducing metals, like fences (which 'bleed capacitance from the land'). We can then install paramagnetic rock, such as basalt, to help focus our intentions, to attract charge and create implosion, to allow rain to fall, he says. Labyrinths, geometric stone arrangements and stone circles can all attract energy and rain. Cultivating blissful states of mind in the inhabitants is also helpful, he believes. Sacred geometry in the form of earthworks might be the go too:

'Bill Witherspoon carves a shallow ditch many acres across in the shape of a Sri Yantra, he then pours in paramagnetic sand and measures a dramatic change in annual rainfall locally'.

[And] '... Marty Cain (labyrinth installer) took a small core of emotionally energised women to a magnetic crux spot in Vukovar, Yugoslavia, shortly after the war-time bombing there caused all the underground water to shrink back from the land surface. She installed a properly dowsed labyrinth and promoted intense healing dialog with the elemental forces. The underground water reappeared.'

Water rituals can also be effective without the need for props. Ceremonies don't have to be long or complicated, either. For instance, on 30th October 2004 I ran a geomancy workshop for the Theosophical Society in Brisbane. In south-east Queensland at that time the drought was bad; a bushfire had been blazing on the nearby Sunshine Coast and its smoke was filling the air.

In the afternoon I took the group of about 20 people for a walk through parkland and down to an artificial waterfall on the edge of the city's central business district. We were highly sensitised from practising dowsing all morning. At the waterfall and pool in a little green glade we tuned in to the lovely water spirit there and honoured her sacred being, chanting in beautiful harmonies. It was a magical time for all. Then I went back to Victoria. The following week I chuckled to see reports of heavy rain and flash flooding in Brisbane. The fires were quenched too. Several people emailed me and suggested – 'Do you think we did it?'

Rain songs re-energise the land!

Richard O'Neill:

'In December 2002, during a long dry spell, the singing group Acappella Round the Bends sang a series of rain-making songs written by myself. The songs were written with new words to well-known tunes, with the intention of enabling people to quickly 'sing-up' and energise themselves and the environment, and it worked! Avalon, the sing-up centre, had the heaviest rainfall in the Sydney metropolitan area ...

In the Kimberleys in June 2005 our group was singing rain songs together. The Aboriginal elders suggested we ought to slow down, or it would bring heavy rain. Sure enough, two days later, on entering the Bungle Bungles, we were greeted by a cleansing storm, a full two months ahead of season. There was so much rain that the Tanami Track was closed for a few days. It has reinforced the theory that singing up the rain really does work.

I like to bridge this sort of traditional wisdom with my scientific background and work with nature's subtle frequencies and songlines. I've been inspired by people like Aboriginal elder, Aunty Beryl Carmichael, who says:'

'My father, the late Jack Kelly was the last rain-maker in the Broken Hill area. Singing and chanting is very powerful. It comes straight from within the soul in our language. We sing up fish, rain and are the custodians of the birthing song of this region.'

References

Faber, Scott, 'Sky rivers – atmospheric water-vapor transport' *Discover*, January 1994, on-line.

'Air-well waters parched farms', *Popular Science*, March 1933.

Nelson, Robert A, '*Air-wells, fog-fences and dew-ponds – methods for recovery of atmospheric humidity*', 2003, on-line.

Whissons windmill at www.waterunlimited.com.au

Adams, Phillip, 'Water from wind', *The Age*, 27th January, 2007.

Browne, Anthony, 'The spray turbine – how an egg-beater at sea could end drought and war', *The Times*, UK, 2nd December 2002.

Flannery, Tim, '*We Are the Weather Makers – the History and Future Impact of Climate Change*', Text Publishing, 2005, Australia.

Cloud Seeding: Farley, Edwina, *The sky's the limit for cloud seeding*, ABC Rural on-line, 11th May 2007. Conference at: www.bom.gov.au/bmrc/basic/events/cloudseeding/CS_Booklet.pdf

McKusick, Robert T. 'Dowsers bring rain to fire', *The American Dowser*, November 1985.

Biodynamics: Lovel, Hugh, *Agricultural Renewal – A Basis for Social Change*, Union Agriculture Institute, USA, 2000.

Bird, Christopher and Tompkins, Peter, *Secrets of the Soil*, Harper and Row, USA, 1989.

Biodynamic Farming and Gardening Association, PO Box 54 Bellingen, 2454, Australia. Email – bdoffice@biodynamics.net.au

Cloud-busting: Don Croft – www.educate-yourself.org/dc/
Trevor James Constable – www.newknowledge.org/ga.php?linkid=31"

Tim Strachan – www.energizewater.com

Dan Winter – www.soulinvitation.com/water

Richard O'Neill – www.spiritsafaris.com

Rain prayers: Kohn, Peter, '*Rabbis urge prayers for rain*', 11th May 2007 at – www.ajn.com.au/news/news.asp?pgID=3208

Coorey, Phillip, 'For millions the water will stop midyear', 20th April 2007, *Sydney Morning Herald*.

'Late rain saves month', *The Weekly Times*, 9th May 2007.

Part Four:
Restoring the waters

4.1 Re-watering Australia's landscapes

Grassroots solutions to soil salinity

Australia's ancient soils are often leached to the bone. Soil fertility is declining with farming methods that are more suited to Europe, and we are simply mining the soil. Chemical-pumped farm soils suffer ever-rising levels of acidity and there is also the growing scourge of soil salinity, chiefly a problem of the south.

According to the Australian Dryland Salinity Assessment of 2000, around 5.7million hectares of Australia's agricultural land, 20,000 kilometres of waterways, 20,000 kilometres of major roads and over 200 towns are at high risk of salinity, *The Bendigo Weekly* reported on 8th April 2004. Huge areas of inland Australia were once an inland sea, so there is a lot of salt in the ground, but it usually stays in check and isn't a problem in the soils of healthy, undisturbed environments. The worst affected is Western Australia, where the loss of arable land was estimated to be 1.8 million hectares in 1998 and possibly reaching 3.3 million hectares by 2020. However, some people, such as Jennifer Marohasy, believe that figures like these have been artificially inflated.

Hardly a new phenomenon, salt was a cause of concern in Western Australia decades ago. In 1917 Professor Paterson, speaking to a Royal Commission on marginal agricultural lands in the south-east, warned that probably one-third of the area was too saline for profitable farming. However, his advice was ignored and the land cleared.

In the early days of white settlement land was given out freely on the condition that it be totally cleared and farmed. The typically poor sandy farm lands were stripped of tree cover and farming methods wore the country down. Many families were forced off when their farms became rabbit-infested dustbowls during drought.

Science at the grass roots

Harry Whittington was born south-east of Perth and the family farm, 'Springhill', had been selected by his father back in 1897. When the old man died in 1942 Whittington took over the reins. In his childhood at Springhill there had been excellent cereal crops, abundant household vegetable gardens, a permanent creek and an ample year-round water supply. However, by the 1930s the creek had filled up with sand, its only water beneath the surface and the seepage that used to run into it from the banks was no more. In the winter wet season the creek was transformed into a muddy, raging torrent. On the slopes erosion gullies, some over 3 metres deep, had gouged the land. Waterholes were either dry or saline, while the valley floor was waterlogged and unusable.

After fifty years of farming, all the fresh surface water had dried up and dams had to be made to catch rainwater. There was also very low fertility. By 1946, with shrinking areas of good pasture, sheep numbers were well down and crop yields from the slopes had gone below profitability. Even the family home was affected by salinity and plaster was falling off the walls. By 1958 it was uninhabitable.

Useless government advice
Harry approached the Department of Agriculture for advice. They told him that the problem was caused by a 'rising water table' and that the land could never be restored to its previous fertility. They advised him to plant salt-tolerant trees, shrubs and grasses in the saltiest areas. Of the many species recommended that he tried out, only some tamarisks survived. The ineffectual advice given to him in 1946 is basically the same given out in 1975, Whittington said in the book he wrote about his experiences.

Whittington saw a film about engineering works done to control soil erosion around dams in the Tennessee Valley, USA, by the Soil Conservation Authority there. They were tackling similar problems to his – eroding slopes and the waterlogging of valley floors. Inspired to study the problem more deeply, he decided to find out for himself if the salt problem really was caused by a 'rising water table', or just by changed seepage patterns. It was around 1950 that he began hand-drilling holes into a salt-scalded area. Some went down to 60 centimetres (two feet), some were 3 metres (10 feet) deep and one went down to 6 metres (20 feet). Below the surface of every hole there was always dry clay. He could never find evidence of a 'rising water table' at all.

Seepage patterns at the top of the slope and near the valley floor were monitored and it was plainly evident that the excess water was coming from sub-surface through-flows. So where was all the salt coming from?

In 1955 Whittington saw data from the Agriculture Deparment showing that rainfall in Perth, coming from clouds formed over the Indian Ocean, was quite salty, at 543 milligrams per litre. The figures were lower for inland areas, with Brookton only receiving 57 milligrams per litre. He worked out that this meant 1.25 kilograms salt to the hectare for every 25 millimetres of rain that fell on Brookton. With an average rainfall of 425 millimetres, this meant that something like 21 kilos of salt per hectare per year was being dumped. With all the land cleared, this salt then washed down the slopes and accumulated on the valley floors, where scalds mostly started. Harry also estimated that 50-60 percent of the superphosphate being applied was getting washed away too.

Yet the Department people would only advocate pasture furrows and the planting of salt-tolerant plants, and the only assistance on offer was help in installing pasture furrows. In desperation Harry agreed to a trial. making furrows about 8 metres apart. But after the first rains they all washed out. These were re-ploughed several times, but after only some 5 millimetres of rain they kept washing out.

First interceptor banks

Finally Whittington set about putting his own ideas into practice and, with a sympathetic bank manager, he was able to get a loan. The Department helped with a little surveying, but they were opposed to his engineering solutions. He made friends with a bulldozer driver and built his first interceptor banks along the contours of his land. These were successful at stopping surface run-off, but not seepage.

He learned that they had to be located starting from near the top of the slope and spaced at close enough intervals down the slope to keep run-off from gaining speed. At approximately every three metres drop in elevation he placed a bank. Banks were designed to collect and hold the average rainfall and were dug down to the depth of the ploughsole/ hardpan. He also filled erosion gullies. 1955 was a very wet year, but the banks held out and did their job, and previously waterlogged areas started to dry out.

Encouraging results

By 1966 improvements were obvious and considerable. Once-waterlogged areas were now dry and supporting pasture. Soil fertility was rebuilding. Surface run-off was now percolating down deep into the sub-soil to recharge underground aquifers and water had begun to again seep from the banks into the creek. In the drought of 1969 this recharged aquifer supplied vital water for stock and drought-proofed the property. People from the Agriculture Department drilled some holes down to 6-8 metres and found good water in the underground reservoir that had gone dry some 40 years before. Salt-scalded areas were much reduced and the wheat crop was greatly improved on smaller amounts of fertiliser.

Despite the high cost of putting in interceptor banks and very active opposition from the 'authorities', other farmers were keen to get in on the act and Whittington began to help others get started. By 1978 enough interest had been raised for a group of keen farmers to begin a formal organisation called WISALTS – Whittington Interceptor Salt Affected Land Treatment Society, after a big public meeting attended by 270 people. They began an educational program with regular field days and held their first school at Mullewa with farmers attending from all over the state. Consultants learn to diagnose problem areas from aerial photos. Membership peaked at 1200 early on and was down to about 300 in 2002, most members living in Western Australia. But still it was still not possible to do collaborative research with the Agriculture Department and all the committee's travelling expenses came out of their own pockets.

Benefits of interceptor banks

Water moves up the slope from an interceptor bank by capillary action, allowing soil bacteria to revitalise the surrounding soil. In one case, monitored with a soil moisture probe, the hardpan (ploughsole) on the scarp had broken up within four years, with an improvement of crop yields of around 250 percent. Non-wettable soils were found to be transformed in a few short years and wind erosion reduced, thanks to denser root material holding soils together.

Sometimes banks must be built on a slight grade to send run-off from a neighbour's property to the nearest watercourse. Banks can also be designed to deliver excess run-off to a dam. But some farmers have used the banks more for drainage than anything else and have been sending salty water into watercourses. This (and sometimes gung-ho and sloppy work) have given the system a bad name in some people's minds.

WISALTS has been involved in community works, flood prevention in towns and conservation work in wetlands. They have also given advice and help to the Main Roads Department and local shires for preventing water damage to roads by locating and sealing through-flows of water.

A dowser's innovations

WISALTS member Laurie Adamson brought two important innovations to the design of effective interceptor banks. Laurie, a water diviner, found that dowsing made the task of finding the sub-surface water through-flows very simple. He developed expertise in dowsing by confirming the presence of the dowsed underground flows with a backhoe, digging holes to check his success rate.

Adamson went on to write a comprehensive consultant's handbook, including the use of 'the wire' for surveying. A WISALTS consultant today can only qualify if he or she is a proficient dowser. Unfortunately, this has only reinforced the 'voodoo' label in the government's eyes!

Adamson also introduced the use of plastic sheeting to seal the banks to improve their effectiveness. This was a great breakthrough, especially where through-flows of water were in sandy seams and banks difficult to seal. He also wrote a booklet for high school students to raise awareness of soil conservation issues.

One man's interceptor banks

Farmer Alex Martin (now retired) has successfully applied the WISALTS methods on his property near Benalla in Victoria. When I visited I could see life returning to a valley that had been badly salt-scalded, thanks to the interceptor banks. Alex has developed his own theories of soil degradation, saying:

'I think the disc plough and rotary hoes create impervious layers and ultimately cause soil salinity problems ... in fact I believe that salinity is caused by three things: 1) poor sub-surface drainage, 2) soil compaction (from livestock, ploughing etc) and 3) soluble chemical fertilisers.'

Martin is a firm believer in getting the organic matter levels up in soil as a water-balancing strategy. The amount of water absorbed by the soil, instead of running off, is hugely increased as organic matter levels rise, he says. To stimulate microbial activity to break down this organic matter, Martin has been applying biodynamic sprays to 300 of his 2000 acres for some 12 years, every autumn and spring. At a field day he learned that the biodynamic spray 500 can help with salt problems too. Getting

The Wisdom of Water

the land's water balance right is the first step, he feels.

'Where has tree planting helped soil salinity? If you put in the banks properly, starting up high on the hill, you help the water to be evenly distributed, rather than waterlogging the valleys. This is the key to controlling salt.'

But the expense of interceptor bank construction has been very high and not accepted as a tax deduction, so there's little incentive for many farmers to take the plunge, Martin points out. Later Pat Dare, of the Permaculture Association of Western Australia, told me that:

'WISALT banks don't always work because people have tried to cut corners and save money and these have given the properly made ones a bad name. People often put banks in too low, instead of starting high up a slope. The low banks can become drains for salt and chemicals going into waterways.'

Afterwards I visited a showcase property owned by the Edwards family at Kweda, an oasis enjoying high yields and attracting good rainfall. Trees have been extensively planted on the tops of hills and interceptor banks were graced by rows of glistening young Oil Mallees – a future eucalyptus oil crop. (You can see the farm on my film *Grassroots Solutions to Soil Salinity*.)

It was exciting to see the wonderful results of getting the water balance right on 'flogged' farmland that had once been reduced to a wasteland; and to confirm that water really is meant to infiltrate into and be stored in the womb of the good Earth.

Photos: Interceptor banks at the Edwards farm, Kweda, W.A.

What has science to offer?

However, it is quite the opposite strategy that continues to be upheld by government policy. Planting trees and using other means to prevent groundwater recharge are the official mantra, while the fingers of our so-called public servants are firmly stuck in their ears! Australia's National Dryland Salinity Program has 'invested in excess of $25 million on more than 50 major research projects since 1993' *The Midland Express* reported on 23rd July 2004. Still, there are few shining examples of salt-affected land reclamation for all their expensive efforts.

Fortunately, not all scientists are blinkered by the unhelpful, unproven soil theories the Agriculture Departments stubbornly cling to. Rob Gourlay, for one, has been quietly developing a scientific methodology which proves that Whittington's assertions of over 50 years ago were correct. Gourlay discovered that gamma ray data can be used to map the electrical conductivity of soils, as well as other soil properties, across landscapes. It is a cheap way of finding out where the salty areas are, and it is always confirmed by the ground truth. With this mapping method one can see that salt will mostly spread via through-flow pathways that are largely determined by geological structures in the landscape. The 'rising water table' theory should now be dead and buried, Gourlay firmly believes, adding:

'There is just no correlation between government models of where salt is expected to be and reality.'

The Wisdom of Water

But public science is still in denial and the millions or billions of dollars of government funding to address salt problems is dependent on following the old 'rising water table' theory. No-one who uses gamma ray data will be accepted for funding and Gourlay has tried to challenge this unfair situation. His research company has won several awards from various government departments and universities for his innovative technology, which is a world first in its practical applications of gamma ray mapping.

Exposing salinity myths
An important contribution to the usually stifled debate about soil and riverine salinity came on 28th May, 2006, on the Channel Nine *Sunday* program. This featured the handful of Australian scientists who cannot accept the scientific dogma on the subject, including Rob Gourlay and ex-CSIRO scientist Maarten Stapper.

In the program Jennifer Marohasy, director of the environmental unit at the Institute of Public Affairs, was critical of the CSIRO's head-in-the-sand stance on reality and its alarmist view of an impending crisis with the Murray River and with dryland salinity generally. She pointed out that levels of river water salinity for Adelaide have halved since salt interception schemes were begun in 1982. Yet the CSIRO put out front-page news that insisted the opposite was the case. When she challenged the claims, the web page was quietly changed and Marohasy concluded that: 'We don't have a salinity crisis, we have an honesty crisis'.

Asked why the CSIRO (Commonwealth Scientific and Industrial Research Organisation) would want to perpetuate the doom-and-gloom scenario, Marohasy suggested that they might have been driven by environmental campaigning and they were also 'concerned about continual funding if they'd fixed the problem.'

Another doubter of the salt theory is Wendy Craik, head of the Murray Darling Basin Committee. Previously she fronted the National Farmers' Federation. In 2000 the NFF had an unlikely team mate in the Australian Conservation Foundation and the two bodies were demanding $39 billion of public money to fix the 'salinity crisis'. However, when questioned on the *Sunday* program, Craik said she was pleased that all that money was not, in the end, forthcoming. Craik conceded that:
    '... *Flawed models had vastly exaggerated the extent of the land salinity threat.*'

In November 2006 a study by Monash University into revegetation of denuded hillsides had found that:
[It is] ...*'not enough to stop rising water tables bringing salt to the surface of surrounding, low lying land ... Solutions that were proposed up to 1990 had now been proved largely invalid.'*

Goulburn Broken Catchment Management Authority CEO Bill O'Kane was in agreement, *The Weekly Times* reported (8th November 2006):
'The idea you can simply concentrate on managing the recharge areas to solve the problem is no longer valid. We are having to rethink many of our strategies.'

In April 2006 a Senate inquiry into salinity had already concluded that salinity policies needed improvement. Both the $1.4 billion National Action Plan on salinity and water quality and the Natural Heritage Trust were recommended to be extended for ten years, after both programs wound up in June 2007. But it was suggested that tougher accountability measures should be conditional and decisions 'based on science and subject to cost-benefit analysis', *The Weekly Times* reported (5th April 2006).

But which science can we trust? Public science is shackled by its keeper, political whim. Industrial science seems bent on destroying the world. Scientific reality is blinkered and public science today seems, to me, a travesty of what it could be. The answer to the problem lies at the grassroots and it's in the hands of the practical people on the land. It is they who must sustain the soil, as have the generations before them, in order for the good Earth to sustain them.

Restoring the land's water balance

To restore the waters of the Earth we need to return the soil's capacity to absorb run-off with soil conditioning, astute planting of vegetation and earthworks. We can help nature to cleanse and store water exactly where it belongs- in the ground. Rob Gourlay also suggests that:
'... in the long term it is best to release treated sewage water into the soil, wetlands or rivers where microbes (as soil and water biology) are far more efficient at filtering and decontamination than humans can be with reverse osmosis or other technologies.'

The Wisdom of Water

In natural landscapes water cleansing is done by nature's 'kidneys' – her wetlands, marshes and bogs. By installing reed beds and wetlands to deal with our wastewater and stormwater run-off we can emulate nature's ways. Wetland expert Nick Romanowski says that some wetland plants can mop up to five times the amount of potassium and phosphorus that they need. In a constructed wetland such plants can be regularly harvested as a rich fertiliser and the nutrients recycled back onto the land.

We can also allow farm dams to become havens of biodiversity, for wildlife to proliferate in. Eco-friendly farm dams can be multi-purpose. They can improve micro-climate and be a handy water storage for emergency use, for fire-fighting or irrigation in dry times.

Beware of siting dams over natural springs, which may cause dam leaks; or the spring may become blocked and not deliver. And give ponds a variety of depths, as much edge as you can, and, if possible, islands. On the dam banks it's best not to plant anything except groundcovers, nor to allow big trees to grow. Cows can trample ponds down over time and turn them into shallow pools full of water plants. These wetlands can be great for wildlife habitat if left as they are, although stock will need to be excluded.

To create wetlands as nature intended them to be in southern Australia (and to control any pesty carp in a drying cycle) Paul Haw says that:
'It is important to fluctuate water levels, especially in the Murray-Darling Basin. We need to have a wetting/drying cycle. In northern Victoria the swamps usually fill in September/October and are almost dry by May ... Far more birds and animals thrive with a wetting and drying cycle, because the soil is able to breathe and grasses get established thereby creating a food chain when flooding re-occurs.'

Shallow ponds are the go, author Nick Romanowski told students in February 2006. Below a depth of half a metre it is too deep for most plants to grow, but it can be good for fish. If you create an ephemeral wetland with plants that are adapted to drying out, this will be like a natural billabong. You'll need to add good soil to the top of the clay pond bottom and shallow shelves can provide planting ledges for water plants, he suggests.

Fish in wetland dams can provide a feed for the landowner. Silver Perch are the best aquaculture fish in Australia. Slow growing but good eating, they can handle a little salinity (up to one-third the strength of seawater.)

Rainbow fish can be a good addition too, as they keep mosquito wrigglers down and are compatible with tadpoles. Silver Perch will not eat tadpoles, but Golden Perch will. Romanowski rests his dams from fish for a year or two, to allow frog numbers to build up.

Duckweeds are some of the many plants that can be grown in wetlands and ponds. They are great for treating high nutrient water, soaking up ammonia, as occurs with fish waste in aquaculture. Ducks will eat duckweed and it makes a good garden mulch too. Your wastewater can thus be recycled and turned into eggs, fruit and vegetables. Now we're talking sustainable!

Re-creating chains of ponds

Ways to re-establish water balance in the landscape typically involve expensive earthworks and construction techniques. But this isn't necessarily always the case, as the work of Peter Andrews attests. Andrews restored the earlier fertility of his own grazing property, Tarwyn Park, a horse stud at the top of the Hunter Valley, after long years of observation and restoration of nature's wisdom. It had been degraded down to 'a salt-ravaged, scoured property ... that couldn't grow anything' around 30 years before. The previous owner had drained the property, and salt was affecting land downstream. But Andrews was eventually able to turn it all around and droughtproof the property as well. He told a crowd, at the Bendigo Eco Expo in 2006, that he was able to achieve this because:
'the natural processes in the landscape are the most efficient to overcome climate extremes'.

Andrews went on to paint a picture of original landscapes before white graziers moved in, when the rivers would run high in the valleys, their frequent flooding providing sedimentation that kept raising the level of the bed and banks ever higher. Beds of grasses and rushes once filled the valley waterways, keeping water flow to a slow movement and thus allowing moisture to seep out into the surrounding areas. Nature's incredibly efficient recycling systems in these swampy meadows ensured that the water was self-cleansing, Andrews explained.

Cones and lenses of water would naturally form in the landscape and these would droughtproof surrounding land. Early explorers described the verdant chains of ponds and their lush environment as a product of abundant in-ground water. Aboriginal dot paintings often depict waterholes as braided streams connecting chains of ponds together, he pointed out. But today, he said:

'Landscapes have lost resilience. For instance – the chains of ponds are often degraded now into incised gullies, where head wall cuts make incisions in the land, from rampant erosion. Then salt leaks out and fresh water is diminished.

'Yet they keep telling us that the water table is a bad thing! When actually it sustains the plants and the land relies on it. We just haven't been understanding it properly ...

'What I have done is to return to nature's ways and recreate those lost freshwater lenses, by making a system of leaky weirs, that slow down water flow and allow it to infiltrate down into the water table ...

'I place logs, rocks and organic plugs at flow points where they can dissipate the energy of the water and make it run higher and slower in the landscape, to leak out to the surrounding soil.'

After several years of drought the Andrews property Tarwyn Park still looked like a lush oasis and eventually he started to gain recognition. A team of investigating scientists from the CSIRO have recognised the value of his achievements. Andrews also told his story on an ABC TV program in 2006 (*Australian Story*) and this gave his approach, which he calls Natural Sequence Farming, more well-deserved exposure.

Saving soil water on the farm

Rob Gourlay:
'Kate and Peter Marshall and their family have created a superb chain of ponds system on their Sunningdale property at Reidsdale, near Braidwood in New South Wales. It has restored the sustainability of the farm and the innovation in this water saving system parallels the success achieved by Peter Andrews.

'The Marshall property and Woodford Lagoon is about 400 acres and was once part of an extensive chain of ponds system that formed the Jembaicumbene Swamp that was first described in the 1840s. When the Marshalls arrived in December 1997 it was heavily eroded with sheet erosion in all areas. There were major gullies on the slopes, no standing water and a deep, down-cut intermittent creek.

They initially dug pits to check the stratigraphy of the soil and get a picture of past erosion and sedimentation cycles (eg. a massive depositional flood in about 1905 that submersed the chain of ponds floodplain and riverine vegetation). They identified old billabongs off the current streamline that were completely filled with eroded soil from upstream. They observed that during rainfall events the local Reidsdale Creek was losing big chunks of stream bank, overtopping banks and stripping away sheets of soil that left the water turbid. During dry times, the creek rapidly dropped and dried out and this process was dehydrating the whole property.

'The family decided to firstly restore soil health prior to restoring the chain of ponds system. They removed stock from the streamline and used a Yeoman's plough to oxygenate the soils and break through the soil hardpan that stopped the water from percolating deep into the soil profile. This increased the soil carbon and biology, and the capacity to store water deeper into the soil structure.

'Then they excavated the sediments that filled the old billabongs and, within a short period of time, created a series of ponds that eventually refilled from increased soil water and flooding. Peter installed small stone weirs and fascines [bundles of wood] in the creek to trap debris and filter water. The family planted the ponds with native rushes, reeds, sedges, nardoo and water lilies. Not only were the chains of ponds restored, the natural lateral flows of water through the soil more enabling deep percolation to hydrate the paddocks back to life as well.

'Restoring chains of ponds is a very cost-effective approach for farmers on these southern tablelands and it can improve soil water storage four-fold, considerably increasing the growing period for plants. Peter says that water storage in the soil is also critical to reducing the destruction of floods and it feeds the creeks that run longer into the dry times. Also, as creek water leaves the property it is cleansed and oxygenated, and the soil nutrient cycle kicks into life again. The property is now alive with frogs, birds and other wildlife that signifies a healthy, living soil and water system.'

Keyline system

The Keyline system, developed over 50 years ago in New South Wales by P. A. Yeoman (1905–1984), aims to increase the field capacity of farmland by allowing it to absorb and hold more water. It is a natural means to prevent soil erosion and salinity and improve soil fertility and plant growth too. Well ahead of his time, Yeoman devised a successful system of sustainable agriculture and was constantly in conflict with the bureaucratic orthodoxy who were (and many still are) opposed to his ideas. While the Keyline system has been taken up with great enthusiasm across the world, no monument yet marks his grave.

The Keyline approach is to harvest rain run-off with dams placed as high in the landscape as possible, rather than letting water collect and saturate the valley floors. Run-off is also channelled to where it is needed via cultivation furrows that follow the contours, using the Yeomans plough. The result is beautiful moist, living soils and lush pastures, increased fertility and water availability and, of course, greater farm productivity. Keyline planning begins with the study of a topographic map of the land, to discover the contour lines that may be harnessed for water-harvesting earthworks and dam placement. Ken Yeoman has continued the farm consulting services that his father pioneered. Based in south-east Queensland, he has fully revised and re-published his father's classic 1978 book on the subject: *Water for Every Farm – Yeomans Keyline Plan*.

Water-friendly ripping and ploughing

To retain moisture in topsoil one of the strategies of Keyline and permaculture design is to deeply rip the ground along the contours when the ground is moist, but not waterlogged. A single tyne deep ripper can be pulled slowly along, ideally ripping to a depth of up to around 500 millimetres (20 inches). And there are other even more eco-friendly implements available, such as the Yeoman's deep ripper, Airway vibrating mole and aerator spike.

Top-dressing the soil with lime, dolomite and/or basalt dust prior to ripping is a good idea too. The oxygen and water that can subsequently infiltrate the ground will allow plants to grow rapidly in the sponge-like soil. Deep ripping is ideally repeated every couple of years.

A major advance in sustainable agriculture has been developed by Victorian farmer Ross Hercott, who has spent many years of research and development into producing a 'revolutionary implement' that can reclaim salty and degraded farmland. The 'Ecoplow' was developed on Geoff Burnside's farm 'Cooinda' at Pyramid Hill in 1993, when Burnside and Hercott began a soil reclamation project on the farm. Soil salinity was a major problem there, along with compaction and hardpanning plus declining rainfall. Hercott says that:

'The original Wallace plough had been around the farming sector since the early 1960s, but required several modifications to bring its design into the 21st century. [For the Ecoplow] ... most of the research and development work has been done in developing a point which cuts through hard ground cleanly and is long lasting in all types of country ... the point [has] the ability to aerate deeply without altering the soil profile.

'...This plough is taking over from where the Wallace plough left off. If used properly, the plough will eradicate many of the farming problems being experienced the world over. At 'Cooinda' now the ground has been transformed using natural farming, deep tillage and no superphosphate.'

Sustainable pastures

Summer active perennial groundcovers, such as lucerne, help to take advantage of any summer rains that may fall and they keep the soil covered and protected. Perennial, deep-rooted forage crops can droughtproof soil and persist for decades. They can provide moist fire-retardant barriers too. Many farmers use their green lucerne paddocks as fire refuges for stock when bushfires come through. Native grasses are also perennial and well adapted to extremes of climate, handling drought and fire, and often re-shooting soon afterwards, thanks to energy reserves in the root systems. Kangaroo grass, for instance, will re-sprout actively after summer rains fall. But stock need to be kept off until it's had a chance to recover its carbohydrate root reserves.

When heavy rain fell on 20th January 2007 in western Victoria there was much sheet erosion, Wimmera farmer Iestyn Hoskins told me in March. Near Hamilton, another farmer, Simon Menzel, said that he also saw a 'lot of brown water rushing off the neighbours paddocks.' But his own paddocks were resilient, with strong grass cover, thanks to a cell block grazing regime. Only clear water drained off them.

'Pasture-cropping' is another way to keep soil moist and healthy. It's a system developed over the past 15 years by Gulgong, New South Wales, farmer Colin Seis, who tackled the problem of soil salinity at his farm Winona in the process. Seis maintains '100 percent ground cover for 100 percent of the time'. Pasture-cropping can be used to grow organic crops without using a plough or destroying existing perennial pasture, and Seis's methods are now catching on in other parts of the world. He says that:

'Pasture cropping is a zero tilling technique of sowing annual cereal crops into living perennial pastures and having these crops grow symbiotically with the existing pastures, with real and advantageous benefits for both the pasture and the crops.

'From a farm economic point of view the potential for good profit is excellent because the cost of growing crops in this manner is a fraction of conventional cropping. The added benefit in a mixed farm situation is that up to six months extra grazing is achieved with this method compared with the loss of grazing due to ground preparation and weed control required in traditional cropping methods.'

Permaculture swales

Hillsides can often benefit from swaling, a type of earthwork that's popular with permaculture enthusiasts. Swales are long, level excavations across sloping land, which intercept run-off and allow water to percolate into the ground. Swales are placed strictly on contour. Otherwise, if they are not level with the contour, they become drains. They can be constructed with deep-ripped incisions, ridges of rocks or strawbales, terrace-like earth mounds or excavated hollows.

Swales provide moist, fertile ground to plant fruit or high-value trees on, usually on the top of the swale or slightly downhill from it. As well as recharging groundwater, swales help prevent erosion by collecting any silt that's running down the slope. Their size and how close they are best spaced apart depends on what the highest rainfall may be and what the soil type is. They can be good on slopes that are not very steep. Swaling can be done with disc mounders or a Yeomans keyline plough, a grader, an excavator or a bulldozer.

Ponding banks in Central Australia

The arid centre of Australia is home to a large pastoral industry which often has a poor record of land management. Here, where the average annual rainfall over the past 20 years has been around 300 millimetres / 12 inches, land can easily become a sea of lifeless sand drift punctuated by rabbits. That was until people like pastoralist Bob Purvis started to restore degraded land by constructing his 'ponding-spreader banks of earth, which slow water flow across the land'.

Many people in the rangelands around the Alice Springs area followed suit and about two-thirds of the district's 80 pastoral properties are now members of the Centralian Land Management Association. They all embrace this same approach to local landcare.

One such member is Dick Cadzow of Mount Riddock, around 200 kilometres north-east of Alice Springs. There was no pasture nor shrubs, just some large trees when the Cadzows arrived in 1986. They immediately began the hard work of reclamation, starting by ripping and cross ripping to remove rabbits and allow rainfall to infiltrate through the hard-packed crust. A scarifier known as the Crocodile was also used; with its prominent ridges on massive wheels the machine was towed behind a bulldozer to leave a trail of pits for water to infiltrate and for plants to naturally establish.

The 2500-square-kilometre property was de-stocked for seven years. Now its 12 main paddocks hold 6000 head of Poll Herefords, which are conservative stocking rates compared to the local average. A rotational grazing regime is used and at any one time a quarter of the paddocks are resting, usually for a year or two.

In 2005 the Cadzows were awarded The Rural Press Landcare Primary Producer Award at the Northern Territory Landcare Awards.
They explained in *The Alice Springs News*, 2nd February, 2006, that:
   'Ponding banks are put in wherever there is erosion, whether sheet erosion or washouts, working from higher in the catchment area to lower. There are hundreds across Mount Riddock, some in every paddock, varying in length and shape. They're about one and a half metres high – any lower and they would erode away. The walls of the banks are sown to Buffel Grass and some ponds can hold up to eight inches [200

millimetres] depth of water after heavy rainfalls. Other banks have been designed to simply spread water more evenly across the land.'

Meanwhile, in the Aboriginal lands of the Anangu Pitjantjatjara and in the Pilbara, Operation Desert Stormwater has also been constructing earth mounds on contours as ponding banks around desert communities. These are improving the environment and trees planted on banks have grown well, dust problems have abated, gardens are watered and flash flooding has been prevented.

It has been a great exercise in water harvesting, often in conjunction with roads and airstrips. Banks are around 400 millimetres high and the ponded areas are arranged in series, for water to overspill into lower areas and then infiltrate into the ground.

Stormwater harvesting in the city

You don't have to live in Australia's arid or rural areas to benefit from stormwater harvesting, as it is a wasted resource just about everywhere. Hard-paved city surfaces stop rain run-off from going into the ground, so cities probably top the list of stormwater wasters. Yet a study of stormwater harvesting systems in 2004 identified only 85 in use or being developed in Australia.

Monash University researchers in Melbourne have developed 'rain gardens' that filter storm water and prevent chemically-tainted urban run-off from polluting the Yarra River, Port Phillip Bay and local streams. The rain garden biofilters are around five square metres in size and they cleanse run-off as it passes through a bed of sandy loam soil growing reeds and other water-friendly plants. They can trap nitrogen, heavy metals and other pollutants, and the water is then used in garden irrigation, or is piped back into the storm water drainage network.

Rain gardens are now operating in new residential developments, including Docklands, and in inner urban areas such as Richmond. And in Sydney, at Olympic Park, the state-of-the-art redevelopment of the site of the 2000 Olympic Games incorporated stormwater collection and treatment by means of wetland features.

References

Paulin, Sally, *Why Salt? Harry Whittington OAM and WISALTS: Community Science in Action*, Indian Ocean Books, Western Australia, 2002.
Adamson, Laurie, *Notes on Soil Conservation for Schools*, WISALTS publication.
WISALTS – c/o president Brian Whittington PO Box 89 Bakers Hill 6562 WA, www.members.westnet.com.au/wisalts/
Rob Gourlay – www.eric.com.au
Duff, Xavier, 'Salt theory discharged', *The Weekly Times*, 8th November 2006.
McKenzie, David, 'Salinity overhaul urged', *The Weekly Times*, 5th April 2006.
Cooke, Stephen, 'Rain plus wisdom equals reward', *The Weekly Times*, 14th February 2007.
Haw, Paul, in Green Connections no. 26, Victoria, 2000.
Finnane, Kieran, 'Reclaiming the land', *Alice Springs News*, 2–3rd February, 2006.
Natural Sequence Farming: 'Of droughts and flooding rains', ABC TV, *Australian Story*, 6th June, 2005.
Butler, Kevin, 'What a ripper', *Farm* magazine May 2007, *The Weekly Times*, Victoria.
Operation Desert Stormwater Harvesting, Bush Tech Briefing #3 at: www.pitcouncil.com.au/1Landmanage/lmframe.htm
'Rain gardens to reduce pollution in Melbourne's waterways', 3rd April 2006, *Monash University News._*
Yeomans, P.A, *Water for Every Farm – Yeomans Keyline Plan*, Keyline Designs, Australia, 1993, www.keyline.com.au
Ecoplow – www.ecoplow.com/product/ecoplow-history.html_
Colin Seis's pasture-cropping: www.winona.net.au/

4.2 Household water conservation

If every home in Australia had a big enough rainwater tank plus water efficient features and grey water recycling, we all wouldn't have to share the pain of the 'water crisis'. Rain tanks come in all shapes and sizes these days. There are even subsidies to encourage us to install them. (Make tanks function by gravity alone, if possible, to avoid the problem of no water during a power failure.)

The reduced-use, re-use and recycling of water in the household is not very hard to do. Those on reticulated mains water can restrict water pressure and embrace more water-friendly technologies such as low-flow appliances. Dry compost toilets should become the norm.

There are many ways to save water in the home and cleanliness fetishes belong back in the 1950s! Dirt can be healthy too! Washing our bodies too frequently robs our skin of its oils, so it isn't particularly healthy anyway. And four-wheel-drive vehicles look more genuine with a good smear of dust.

Low-water gardening

Xeriscape your garden! This means that it will run on very low water requirements. You may never need to water, apart from the establishment phase. So this could be a garden of true liberation, with freedom from watering and from the smelly, noisy ritual of lawn mowing, an assault on the ears each weekend. Is it really possible? Yes! But people in Australia will have to get over their lawn addiction. There are drought-hardy indigenous grasses and ground covers to replace lawns with. Kangaroo grass, for example, is an attractive perennial species with gorgeous purple coloured stems and copper brown seed heads.

English country gardens and endless lawns belong in drizzly old England. Bring on the dry native and Mediterranean garden instead! And the desert cactii and succulents. A word of warning though: spiky plants are not considered very good feng shui, while prickle-free succulents often provide juicy meals to possums!

If you have a vegetable plot, keep it in its own high water use zone. In fact, grouping your garden into 'guilds' of plants with similar watering needs is the way to go. Then nothing gets over or under watered, especially if there is a timer on your tap. Provide windbreaks and shelterbelts to exclude wind and keep down evaporation rates.

When selecting plants consider the amount of rainfall where they originate from. You can always select plants from drier areas during a drought. Plant them on slopes and mounds where good drainage will keep them happy if there is heavy rain. When planting, consider the use of wetting agents if you have non-wettable soils. And planting with a few handfuls of jelly-like 'water crystals' mixed into the soil will give new plants slow-release water surety for over a season or more.

Use lots of mulch for high moisture retention in soil, as up to 70 percent of garden water can be saved from evaporation this way. Mulch also keeps the soil cool in summer and is therefore hospitable to beneficial soil life. However, in winter, when soils need to be warmed up, mulch may be counter-productive. Then you may have to peel back the mulch layer and save it for later or compost it.

You can harvest stormwater in your water-friendly garden by allowing it to soak into the ground, or channel it into a wetland or reedpond. Small frog-friendly, natural ponds are a good addition to your garden too. (To keep out frog predators ponds may need to be netted.) Create a curvy, rocky creek bed to cleanse and energise flowing water and enjoy the high-negative ion environment.

Grey water in the garden

The plants in my little kitchen garden became stunted over the course of last summer. Too much grey water from the kitchen sink was probably the cause. They have only perked up and actually grown since good autumn rains have been falling.

'Grey water' is all household waste water, except for what comes from the toilet – that's 'black water'. We do need to be careful about how we use grey water in the garden. It can be fairly greasy and can clog the pores in the soil, if you are repeatedly watering in the same spot. Concentrated detergents and the like can be poison to plants, especially young ones.

Source household cleansers with low or zero phosphorus levels; simply stating 'biodegradable' on the label isn't good enough!

Rotate garden areas where grey water is used and be aware that it's recommended to avoid watering food plants with it. You can filter it through a grease trap before using it and this can be as simple as a wire basket half filled with straw that is regularly replaced. Or filter grey water through a small treatment pond or reed bed, although this is more practical for a cluster of homes. Another way to use your grey water is to soak your compost heap or worm farm with it. Although local laws differ, it's best to say that grey water is a no-no for sprinkler use. You could instead use drippers or soaker hoses laid under mulch, but some type of initial filtering of the water would be needed.

You can also safely apply grey water over thickly mulched areas and 'no-dig' gardens. It's best to do this on flat ground. Never over-water with it, or water when it is raining. Only use whatever can be absorbed in one spot and don't allow any run off, especially away from your property.

Always use grey water fresh. Don't store it for more than 24 hours, and try not to splash it on the leaves of your plants. Wash your own hands well after using it. And don't water plants at all if they don't need it!

Urine in the garden

Toilet flushing uses a high proportion of urban water and much in the way of savings can be achieved when our toilet waste is kept separated. A review of water recycling by the Australian Academy of Technological Sciences and Engineering in 2004 found that:

'Urine-separation not only allows the collection of 50-85 percent of the major plant nutrients currently present in sewage, within less than one percent of the wastewater volume ... but it also reduces the odour and pollution (leachate) issues associated with collecting/composting human faeces.'

If you are a gardener you'll know that soil needs regularly feeding, particularly if intensive food production is the goal. The production and long-haul transport of commercial fertiliser usually involves a high degree of unsustainability. As an ethical gardener you might ask: 'Where is the most sustainable source of nitrogen to use on plants?'

Well, search no longer. Just stop flushing urine down the toilet and pee in a special bucket instead (ideally with a lid and spout). Add other grey or waste water to dilute it (not too much grey water though) or top it up with some rainwater from your tank. By evening you will have collected a wicked brew of fertiliser soup to pep up plants with.

Water with it carefully around root zones, trying to avoid getting it on leaves. Vegetables, in particular, will love it! Potatoes will go berserk! So will onions, peppers, celery and carrots and anything else that loves a good dose of potassium, phosphorus and all the other juicy minerals that come with the urine.

A dilution of up to five parts of water to one of urine can be used, although native Australian plants may well prefer it weaker (certainly they tend to be touchy about phosphorus). Don't splash it on any leaves or vegetables that you are about to eat. Little seedlings would prefer to be watered with a weaker dilution at, perhaps, ten to one. Always use your brew fairly fresh or it will start to pong. And give plants this weak urine solution often. Daily should be okay, for instant green thumbs!

With an output of some 1–3 litres of urine per day, it is estimated that the average adult produces enough nitrogen in their urine to fertilise around 300 square metres of garden (applied at the annual rate of 70 kilos of nitrogen per acre/0.4 hectare.) So this is both a good financial saving, compared to buying in fertiliser, as well as an environmental win.

Another great way to use urine is to compost it first. You can regularly drench your compost heap with it, sprinkling it in a dilute form. If there are a lot of materials that are slow to break down (such as bundles of weeds, shredded paper or cardboard, straw etc), your wee tea will really give the composting process a kick start! Another method is to half fill a 20-litre bucket with sawdust as your wee container. Sawdust needs a good source of nitrogen to break down. When it's damp and starting to smell, tip it on your compost heap as a thin layer, cover it with over with straw or weeds, and start a fresh bucket up. Your compost will be like dynamite so expect explosive plant growth!

Is it safe?
Pathogen levels in urine are generally negligible, certainly when compared to those in faeces. Urine is even consumed by people as urine therapy or when lost in the desert, so it can't be too bad. However in terms of local planning laws, it's a bit of a grey area. It's probably best to avoid discussing using urine with environmental health inspectors. What does the Australian Academy of Technological Sciences and Engineering think? They suggest dedicated pipelines to take urine from the cities back to the agricultural areas to replace industrial fertilisers. They also state that:
 'Stored urine is hygenically safe and is cost affective for local utilisation.'

Who doesn't like your urine?
Worms in worm farms and vermicompost heaps really don't like to take a 'golden shower' and dry compost toilets are preferably kept urine-free too. However if we separate out the urine and use it in a positive way, it's a win-win solution to environmental pollution.

People who are sick (e.g. with urinary tract infection) or taking medications may not want to use their urine, in case of unwanted side-effects. But if properly composted first this shouldn't be a problem, as any pathogens should be broken down in the process.

Because of its high salt and acid loads, undiluted urine can burn and kill the leaves of your plants. (This is why there is a country tradition of men pee-ing *underneath* the lemon tree, but never on it.) This deadly attribute can be used to advantage to get rid of weeds, if you spray urine onto leaves several times, and on hot, sunny days.

Full-strength urine can be sprayed on fruit trees when they are dormant, to protect them against mildew, apple scab and fungal diseases. I've read that urine is also a good deterrent against marauding deer. A few pails of urine placed in the corners and middle of the garden do the trick, apparently. (This could be tried on other pests too.)

Apart from pests, your garden will love you, and so will Mother Earth too, if you always remember that:
 '*Wee will fertilise!*'

Alanna Moore

Sustainable low-water diet

To save the Earth and its water we need to cap growth and greed. On an individual level each of us can adapt our consumption patterns and match them better to the land's capacity to sustain us. With the irrigation of crops and pastures accounting for around three-quarters of global water use, this is a significant area where water use reductions could be made. To do that, we would have to change our diet to omit water-guzzling food crops.

It's a shock to discover that it takes 11,000 litres of water just to produce one hamburger! So to be truly serious about reducing water demand we need to greatly reduce, or eliminate totally, meat and dairy products from our diet. So much grain is produced to feed livestock (who take up so much space) that if the world went vegetarian, there should be no water crisis or food shortages.

We can achieve excellent health and vitality with a good vegetarian or vegan diet. And it's even better for our health, and the planet's, if we can produce our own food in our gardens with the waters that we harvest on-site – that blessed milk of Mother Earth.

Milk sucks water

Animal Liberation Victoria recently took up a new campaign, with plans to distribute information packs to secondary students describing the animal cruelty and environmental degradation that can result from dairy farming. (Victoria's dairy industry accounts for 80 percent of Australia's dairy exports and about 11 percent of global dairy trade.) Their new campaign also challenges the dubious health benefits of milk and the extreme amount of water involved in milk production.

However dairy-farmer-group spokesman Doug Chant tried to set them straight (in *The Weekly Times*, 16th May, 2007), explaining:
   'We are not the biggest users of water. We use 17 percent of agricultural water, putting us behind livestock, grains and cotton.'

If milk production was greatly reduced, perhaps dairy farmers might turn to producing other, more water-friendly crops, such as hemp for fibre – a perfect replacement for unsustainable cotton crops.

References

Smith, Keith, *The Australian Organic Gardener's Handbook*, Lothian, 1993.
Hunter, Beatrice Trum, *Gardening without Poisons*, Berkeley, USA, 1964.
The Australian Academy of Technological Sciences and Engineering, *Water Recycling in Australia*, a review, 2004.
Smart Gardens for a Dry Climate. Free booklet online at: www.coliban.com.au/smart_gardens
Lunghusen, Felicity, 'School's in on dairy protest', *The Weekly Times*, 16th May, 2007.

4.3 Improving water quality

Wherever water is sourced from there are potential contamination issues and whatever we do to water we typically downgrade its purity and energetic quality. Which water to drink? We need to have a handle on what there is in our drinking water to know if it is going to be healthy or how to improve it.

The reticulation system may be filthy and mains water from house taps can be loaded with heavy metals from rusty iron or old lead pipes; dam water may harbour creatures large and small; and drinking recycled water can put us all on the contraceptive pill!

Contaminated river waters

In many parts of the world treated sewage water is put back into the waterways only to be used again by the people downstream. It's a problem for people living downstream on rivers such as the Mississippi. New Orleans cops a toxic cocktail, as do the cities of Europe's Rhine River Valley. And Adelaide sources up to two thirds of its water from the Murray River, after many towns (including Canberra) have discharged their waste water into its tributaries.

If water treatment plants break down and spillages of untreated water occur, contaminated water can spawn disease outbreaks which can sometimes be fatal to the young, elderly and frail. The treated wastewater ideally goes through a tertiary process to strip it of all contaminants. But even though substances may be removed physically, their energy patterns can remain in water.

Disinfection of water produces its own hazards. When chlorine reacts with naturally occurring organic matter it can produce carcinogenic trihalaomethanes and haoacetic acids. In Europe public water chlorination is being phased out in favour of less harmful alternatives.

The glow in your glass

Possibly the worst-case scenario of all arises when nuclear power plants discharge contaminated water into rivers which people downstream then drink (although it is not an issue in Australia). Anti-nuclear campaigner Helen Caldicott tells us that the water used to cool a nuclear reactor core becomes heavily contaminated with tritium (with a radioactive life of 200 years) or radioactive hydrogen and carbon$_{14}$ (with a life of 114,600 years). You would think that this would be treated as hazardous waste and contained safely somewhere. But instead, says Caldicott, in her 2006 book that won the Australian Peace Prize:

'... *it is routinely and blithely discharged into seas, rivers or lakes, from which people obtain their drinking water ... The true and energetic and economic costs of nuclear power are presently grossly underestimated.*'

Endocrine-disrupting chemicals

Back in the late 1990s, the shocking report came out that up to half of the male fish in England's lowland streams were showing signs of feminisation. Roach fish in ten lowland rivers, downstream of sewage treatment works, had a high proportion that looked as if they had been exposed to oestrogen. They were even developing eggs in their testes, said Susan Jobling of Brunel University, an expert on the condition known as 'intersex', on the ABC Radio National program *Earth Beat* (17th January 2004).

The fish had been affected by a broad range of chemicals that are collectively known as endocrine-disrupting chemicals. EDCs are chemicals that either mimic or interfere with natural hormones that control animal growth and development – mainly oestrogen and testosterone. EDCs can trigger unwanted endocrine changes at extremely tiny concentrations.

The end result of this is diminishing fertility rates for aquatic wildlife and probably for the humans drinking that water too, although perhaps that wouldn't be such a bad thing. Some people also link EDCs to higher rates of breast and testicular cancer, early onset of puberty, and, possibly, to increasing rates of endometriosis and ADD (attention deficit disorder), they said on *Earth Beat*.

EDCs include pesticides, organochlorides such as PCBs and dioxins, alkyl phenols and heavy metals. Then there are the natural and synthetic hormones, such as are in the contraceptive pill, which is excreted fairly unchanged and is difficult to break down. As well as pharmaceutical drugs, detergents, insect repellents and toilet cleansers, EDCs are also associated with animal and mining waste, pulp mills, veterinary drugs and the like.

Tertiary treatment of drinking water sourced from rivers is much more common in Europe than in Australia. In 2002 the Global Water Research Coalition listed endrocrine disruption as a number one research priority. Australia's CSIRO Land and Water division have been assessing the EDC risk in the Murray-Darling River Basin, with Rai Kookana as their project leader. In 2004 Kookana said that a lot more hazard assessment needed to be done and that:

'... there are many chemicals and naturally produced hormones that are detected in sewage at very low concentrations. There is evidence of effects on populations of shellfish, fish, frogs and alligators, where partially treated sewage is discharged into surface waters.'

Where river deltas meet the sea

It's lucky for most Australians that we live around the edges of the continent where rivers carry pollution loads quickly out to discharge into the marine environment. It is not so lucky for the sea waters though. Commercial fishing in Sydney Harbour was banned back in January 2006 because of high dioxin levels found in seafood, a result of the pollution load there. Oceans that have been fertilised with nitrogen and phosphorus become infested with algae and bacteria that can suck most of the oxygen from the water, creating 'dead zones'. There are an estimated 200 'dead zones' around the world and this is up from 150 just two years ago, the United Nations Environment Program recently warned.

Some of these zones are inhabited by stinking, toxic slime. Queensland's Moreton Bay has fallen prey to such a pestilence and fishermen there risk painful skin lesions if body contact is made. Locally called fireweed, *Lyngbya* is a strain of cyanobacteria, an ancestor of modern bacteria and algae, the *Los Angeles Times* reported (30th July 2006). Thirty wastewater treatment plants discharge water into Moreton Bay.

Who wants to use recycled water?

The push is on to use more recycled water in Australia. Already trial systems have been operating for a number of years with some agricultural and forestry enterprises doing well. However, a referendum in 2006 in Toowoomba, in drought ravaged south-east Queensland, received a massive 'no' to the idea, despite the dire water shortage.

While it may offer a lifeline in a drought, not enough information is known about the safety of drinking the treated water. And people's intuitive fears may be well founded, thinks Rob Gourlay:
'I don't blame people for avoiding the decision to consume recycled sewage water as the science in testing all contaminants in water is not adequate ... Reverse-osmosis treated water would be at the top of the list of water to avoid because it will allow about 60 percent of hormones to filter through. A very fine nano-filtering system can remove these, but it's very energy intensive to push the water through it, therefore a costly cleansing method.'

Drinking of recycled water should be only a 'last resort', Australian water expert Professor Don Bursill said in *The Age* (5th June 2007). Bursill told a Senate inquiry about an international study of 98 waterborne disease outbreaks, including one in Canada where seven people died. It found that the technology in water recycling plants was largely okay, but that it was human error in operating them that had caused 80 percent of the outbreaks. There can be no guarantee of complete safety for water sourced from sewage, Bursill believes.

Back in 2006, recycled water hadn't been long on the horizon before lettuce growers in Werribee were dismayed to find their crops suddenly become stunted and sick. A water board investigation into the quality of the recycled water they had been irrigated with failed to find a cause for their death. But in April 2007 the vegetable growers released their own study, which found that the most likely cause was high levels of chlorine in the recycled water (although Melbourne Water denies this was the case). Understandably, the Werribee vegetable growers are not keen to continue using the water! A grower spokesperson said in *The Weekly Times* (18th April 2007) that, in all probability, it was over-chlorination that had:
'caused a drop in pH, which stressed crops and led them to absorb excessive levels of iron ... Their symptoms were typical of iron toxicity.'

Chemical cocktail, anyone?

Many Australian cities and towns have compulsory fluoridation of water supplies for its supposed dental benefits, but most European countries oppose this practice. Fluoride is a highly toxic waste product from the aluminium mining industry. As a compulsory medication of unproven and dubious benefit, it's quite an outrageous imposition!

Brisbane rejected water fluoridation. A 1987 study by Mark Deisendorf, of the Australian National University, looked at rates of tooth decay among 10 year olds in Brisbane and Melbourne. Even though Brisbane didn't have fluoridated water while Melbourne did, the Brisbane kids' teeth were slightly healthier.

Some unpleasant health problems associated with fluoride ingestion include dental and skeletal fluorosis (which is rampant in India and China where fluoride in groundwater is very high), kidney damage, infertility, tumours, hyperthyroidism, increased aluminium and lead uptake and reduced magnesium uptake. As well as problems from ingesting chlorine and its by-products, there's also the added potential harm from inhaling chloroform gas emitted from chlorinated water during a shower or bath.

Municipal water authorities often use aluminium sulphate as a flocculent, to make water look clearer. Sydney eventually stopped adding it a few years ago, after investigation into the health risks. Aluminium accumulation in our bodies is associated with several disorders, including Alzheimers disease, which has only become epidemic in the Western world since aluminium cooking vessels have been mass produced following the end of World War II.

Testing the waters

In times past, new sources of water for the Maoris in volcanic areas of New Zealand might well have been deadly poisonous. Some mineral-rich hot spring waters there run yellow, green, blue and brownish red. Water has always had to be assessed in case it is 'waimate' – of bad quality. Most of the waters in the landscape were considered good – 'waimaori'. Information about water quality was recorded by Maori in traditional songs that were handed down through the generations.

> Nowadays we can test our drinking water for contaminants in a laboratory, or test it ourselves with a kit. Australian school children test water from creeks and rivers for its biological quality, measuring levels of turbidity, oxygenation and other factors using Waterwatch and Streamwatch kits.

Groundwater contaminants

Shallow underground sources of water in southern Australia are often found to be saline. Salty bore water can spell fast death for metal rain tanks. Plastic (polyethylene) tanks are more resistant to this problem, although PVC tank liners may not be guaranteed to last long if they are attacked by microbes in bore water.

Bore waters in Bangladesh and parts of India often contain traces of more worrying contaminants, such as arsenic. (Arsenic also contaminates areas in Australia where gold mining or stock pesticide dipping has occurred in the past.) In agricultural areas, nitrate contamination of ground water supplies is often rife. Nitrate is a natural by-product of decaying sewage, garbage, animal manures, and fertilisers such as potassium nitrate and ammonium nitrate. Highly soluble and mobile in the soil, nitrates easily leach down into the ground. Shallow, poorly sealed or constructed wells, and wells that draw from shallow aquifers, are at greatest risk of nitrate contamination.

Considered a serious health hazard for people, nitrate fertilisers aren't much good for soil and livestock health either. Drinking water with high nitrate levels can cause methemoglobinemia, or 'blue baby syndrome', a condition of impaired oxygen transportation in the blood, to which children are particularly susceptible. A single exposure can affect a person's health, and nitrate is a suspected carcinogen. There are much higher rates of stomach cancer fatalities where high-nitrate water is drunk.

Reduced till, ridge or no-tillage methods, plus crop rotation are some ways to slow nitrate leaching from the soil into groundwater. But low-till is also associated with increased chemical use. Changing to organic farming is probably the best way to avoid the problem.

High levels of minerals, such as calcium or iron, should also be avoided in groundwater for drinking. One really doesn't need to risk mineral imbalances in this way. Iron can be filtered out, but iron bacteria make water brown, unsightly and in need of sterilisation, as does water with any other bacteria present.

Acid rain

Many people have the romantic notion that rainwater is 'pure', when actually it can often be far from pure. Acid rain, for one, hasn't gone away, although improved technologies have reduced air pollution somewhat.

Because rain absorbs CO_2 in the atmosphere it becomes slightly acidic, with a pH of around 5.3. Add in air pollution and exposed stone buildings or ancient petroglyphs and metals will corrode faster. There are increasing human health problems involving eye irritation, asthma and bronchitis which can be attributed to this problem.

Acid rain is created when nitrous oxides and sulphuric oxides are released naturally by bushfires and volcanic eruptions. However, much more is created these days by human activity. The main problem is the burning of fossil fuels such as petrol and coal. The problem has been partly alleviated in Australia since 1985, when cars were all fitted with catalytic converters on their exhaust systems and now emit less nitric oxide. But there are more cars than ever on the roads these days.

The other major source is coal burning for electric power stations. Coal is Australia's biggest export, eclipsing all others. Some coals, such as that mined at Leigh Creek, South Australia, also have high levels of sulphur. Victoria's 'dirty brown' coals are notorious polluters, with the highest levels of carbon dioxide emissions. Mercury is another contaminant released from coal into the atmosphere by power stations.

Acid rain can directly affect the plant life it falls on, and in soils it can mobilise toxic aluminium and thus hinder plant growth from the roots up. Some soils are more affected than others – granite soils more so than limestone soils, for instance. Broadcasting crushed basalt dust ('blue metal'), which is highly alkaline, over affected acidic soils can be a valuable way to remineralise and improve fertility of the soil. (My book *Stone Age Farming* delves into this.)

Water storage and conveyance

Putting pure water into unhealthy storage containers degrades its value. Just think of the Romans whose use of lead pipes, plates and utensils, made them mentally deranged!

To help keep water's integrity, it needs to be kept cool and in a dark place, especially when it is stored in plastic. Plasticisers can leach from containers and pipes. It's recommended not to leave soft plastic bottles of drinking water to bake in a hot place nor to bend them, or even re-use them. The big plastic water tanks themselves will start to break down after a decade or two anyway, attacked by ultra-violet radiation. Keep them as shaded as possible to lengthen their life.

A study in northern New South Wales in the 1980s also found that polypipe, a standard for water conveyance, leaches lead and cadmium into water in its first few years of use. This is worse in hot conditions, so always bury pipes underground and you could always flush them first, only using the water for non-potable purposes.

Schauberger said that water should not be piped through lime, cement or metal; rather the better conveyance is in wood, bamboo or stone pipes. Glass bottles are probably the best medium to store drinking water in and they allow it to keep some of its energetic integrity. Ceramic containers can be good too, but watch out for lead glazes!

Collecting rainwater

Dirty gutters and roofs with bird and bat poo, rotting leaves and the like, are common pollutants of tank water. Dust from contaminated soils from mining, agriculture or industrial sites can blow onto roof-tops and potentially deliver a toxic harvest. Aerial crop spraying can make rainwater lethal, and in cotton-growing areas one never drinks it! Tar-based roof coatings can bind organic chemicals, such as pesticides, to a roof surface and are not recommended for the collection of rainwater; neither are tiled roofs suitable.

Inside tanks algae can grow if it's not dark enough, and if creatures can get inside, there may be the odd dead animal lurking and frog poo too. If mosquitoes can get in and breed up there's the added danger of people contracting mosquito-borne viral diseases.

Heavy metals from household plumbing, old peeling lead paint, and metal roofs and tanks can be a problem in tank water. New zinc tanks give an unpleasant metallic taste to the water. If there is more than one sort of metal in your roof, gutters and spouting, then the interacting metals will leach more rapidly into the water. An even worse problem emerged in a case where the rooftop air conditioner in a home was found to be faulty and the whole iron roof became electrically live as a result. Jim Williams, writing in *Green Connections* magazine (no. 21, 1999) said that:

'The current from the roof had been conducted along the galvanised downpipe to the rainwater tank and electrolytically dissolved the zinc tank lining into the water!'

Creosote, as fallout from your wood-fire chimney or flue, is another potential contaminant and a known carcinogen. Government guidelines suggest not collecting rainwater from parts of a roof where any chimney flues may be located. But this is kept quiet by wood heater makers and retailers. What is the level of risk? There's not a lot of information out there. However, my local chimney sweep tells me that a good annual clean-out of my flues should avert the problem.

To counter contamination from the roof, devices that divert the first flush of rain are available. Regular checking and cleaning of gutters is a must. A filtering system in the home could be worthwhile extra insurance for good water quality too. Without regular cleaning, filtering and care, sludgy debris will soon start to accumulate in the bottom of your water tank and tanks will need cleaning out every few years. When concrete water tanks are installed underground, leaking septic tanks have been known to contaminate the water in them. New concrete tanks can leach lime and impart elevated pH levels to water.

The 'water crisis' is partly resolved by the addition of water tanks to every home (including first flush diverters). Why has this not been done already? It turns out that by-laws have long banned water tanks in urban areas for a range of reasons, from the aesthetic to the plainly paranoid (such as concern about water leaking from poorly installed tanks). Rob

Gourlay alludes to the true reasons for the reticence to allow tanks when he says that:
'The water utility companies want us to use the current reticulated infrastructure to provide the capital for future maintenance costs and new infrastructure ventures like desalinisation... There is a need to update plumbing regulations to allow houses to implement the recycling of grey-water and change health regulations to allow people to collect rainwater in household tanks.'

The Australian Conservation Foundation couldn't agree more. They announced in April 2007, at the conclusion of a study they had just undertaken, that:
'Rainwater tanks are a cost-effective solution to the urban water problems plaguing Melbourne, Sydney and South-East Queensland... Rainwater tanks are five times more energy efficient than desalination plants and twice as energy efficient as the proposed Traveston dam proposed for Queensland's Mary River ...
'While 38 percent of households in Adelaide have rainwater tanks, fewer than 6 percent of the houses in Melbourne, Sydney, South-East Queensland and Perth do.'
'If governments systematically installed rainwater tanks in Australia's major cities, we would secure as much water as the planned Kurnell desalination plant in Sydney, the Tugan desalination plant on the Gold Coast and the stage one of the unpopular Traveston Dam.'

Acid or alkaline water?
As for the pH of water – it's a confusing and hyped-up subject. Pure water is neutral at a pH of seven. However, some people selling water treatment products claim that alkaline water is what everyone should be drinking. But considering the acid state of our stomachs and the fact that we all individuals, some people are now questioning the veracity of these claims. In fact, there is no agreement as to what the ideal pH of water should be for us, writes Dr Jaroslav Boulbik (*Options*, issue 9, 2006), adding that:
'There are some studies to suggest that the consumption of alkaline water may have significantly deleterious effects, such as degradation of heart muscle and effects on cardiac enzyme levels [in a study on rats] ... On the other hand acidic water has been shown to be useful topically to manage skin diseases and counteract the effect of alkalising detergent and soap products.'

Filtering and sterilising water

If you want to improve your water quality it's probably best to start by filtering out any possible contaminants first. The type of filter will depend of what kind of pollution is present. In pre-colonial western Victoria the Wotjabaluk people made possum skin bag containers for water. The fur was left on the inside to filter out any dirt from the drinking water.

These days carbon filters are a cheap and popular way to remove many unwanted contaminants, but reverse osmosis filters are needed for the more intractable ones, such as fluoride and many, but not all, EDCs. Boiling or letting water sit before drinking will allow chlorine to outgas. But boiling water with nitrate in it will only make it worse.

After filtration there may be a need for sterilisation. Microbes can be removed with ultra-violet light, which is very effective and does not produce the disinfection by-products that chlorination produces.

Ozonation is another great alternative, and it is more effective against cryptosporidium than chlorination. Many European and American cities purify their public drinking water this way, although The Australian Academy of Technological Sciences and Engineering note that ozone treatment is not always easy to use consistently.

Oxygenated water

Oxygenated water has long been used in hydrotherapy, which is said to be the oldest healing modality on the planet. Ed McCabe in America promotes ozone charged drinking water as a therapeutic aid for many diseases; while 'thousands of doctors in Europe prescribe it to patients', he says. The O_3 gas is put into water by bubbling it in from a machine. It must be imbibed immediately, while the O_3 is still in the glass.

In a similar vein to McCabe, an author in 1950 admonishes the drinking of well or bore water, calling it the 'deep sunless waters', warning that it causes a loss in vital energy, disease resistance, powers of self-restoration and self-purification. In her book *Healing by Water, or Drinking Sunlight and Oxygen,* one reads that rainwater, containing more oxygen and having the 'power of absorbing the rays of sunlight' is preferable, and that this 'living water' was known to be therapeutic since ancient times.

> Victor Schauberger, the Austrian 'water wizard', also didn't approve of drinking groundwater that hadn't emerged naturally as a spring. When water flows along the surface in natural curves, eddies and whirlpools, and past beneficial rocks, it becomes oxygenated in the process. He called well and bore water 'immature' or 'juvenile'.

Solar water purifier

John Ward, a retired South Australian physicist, was holidaying in Zimbawe when he saw the misery of people who only had contaminated water to drink. He resolved to find a solution with low technology, but it wasn't easy. Eventually he designed a device that has no moving parts, no electronics, no costly filters and uses no chemicals. It converts many types of contaminated water, including sea water, into healthy drinking water.

The unit looks a bit like a solar hot water panel, with a special sheet of glass on top that maximises the sterilising ability of ultra-violet light and the heating ability of the infra-red rays needed to evaporate the water. Impure water or seawater enters via an inlet at the top of a cascading tray.

As the sun shines through the glass into the water, heat is partially absorbed by the water and then more completely by the black plastic panel of cells. The water vaporises and condenses on the underside surface of the glass and then runs down into the bottom of the unit.

The water by then is purified and sterilised to a level better than standards set by the World Health Organisation and the unit has been:
   'recognised as the worlds most advanced and efficient solar evaporation water purification system.'

Rotary International's 'Safe Water Saves Lives' committee, who had been searching the world over for the best solution for water purification, realised their search was over when they found The Solar Water Purifier™. They are now funding its local construction and are distributing units in many countries around the world.

Desalination and wave energy

Desalination is often given as an answer to Australia's lack of water. After all, seawater is a vast resource and most of the population of the world (some 60 percent) live within 60 kilometres of the coast. But ex-New South Wales premier Bob Carr famously called desalinated water 'bottled electricity', because of the high energy costs involved in normal desalination plants. Fortunately it is the very energy of the waves being desalinated that can power this usually expensive process. Waves are predictable, unlike wind or sun, so it's easy to match supply and demand.

Stephen Salter (of 'Edinburgh Duck' fame) and his team at Edinburgh University are working to perfect the small-scale device to desalinate seawater. Their Ducks can convert wave energy into pressure changes that aid the collection of pure water as steam from seawater. The Ducks operate in a piston-like motion, slowly but surely creating salt-free water that is pumped back to shore through the two 'legs' that tether the Duck to the seabed. When finalised, units of 10 metres by 20 metres in size could be large enough to supply water for 'more than 20,000 people', says Salter in *New Scientist* (7th November 2006).

Now an inventor in Western Australia, Alan Burns, is developing an even more environmentally friendly wave-powered desalinator. A world-first base-load renewable energy power station and zero emission desalination plant, the device is named after the Greek ocean goddess. (Ceto was 'probably the Syrian fish mother Derceto incorporated into Hellenic legend,' says Pat Monaghan.) CETO sits safely on the sea-floor, where:

...' *An array of submerged buoys is tethered to seabed pump units. The buoys move in harmony with the motion of the passing waves, driving the pumps which in turn pressurise seawater and deliver it ashore via a pipeline, where it can be used to supply a reverse osmosis desalination plant ...*

'*The high-pressure seawater can also be used to drive hydro turbines, generating zero-emission electricity ... Each unit can produce enough power for about 50 homes ... [And] all of Australia's southern mainland cities' current water needs could be satisfied by CETO units covering an area of 155 hectares (70 football ovals) of sea floor at around 75% of the price of current desalination projects.*'

New wave of water conditioners

Rob Gourlay:
'Some great work on conditioning damaged water has been undertaken using combinations of electro-magnetic energy, vortices and microbial treatments. Water conditioners are now in common use for changing damaged water structure (hydrogen bonds) and minimising the effects of saline water, etc.

Also, Effective Micro-organisms (EM) were developed in Japan during the 1980s to treat surface and sub-surface waters and this technology shows great promise of cleaning damaged water and improving soil health. That is, the right populations of microbes within a soil or in flowing river waters are much smarter at water conditioning than human engineering with membranes, electrolytes, ultraviolet and other chemical or engineering technologies.'

Flowforms

An elegantly sophisticated way to energise and purify water is to run it through a cascading series of flowforms, which can be bought ready for use, or as mould kits. Flowforms allow water to become well oxygenated and are used for all types of water conditioning, from attractive water park features to dairy effluent cleansers and for biodynamic stirring of preparations.

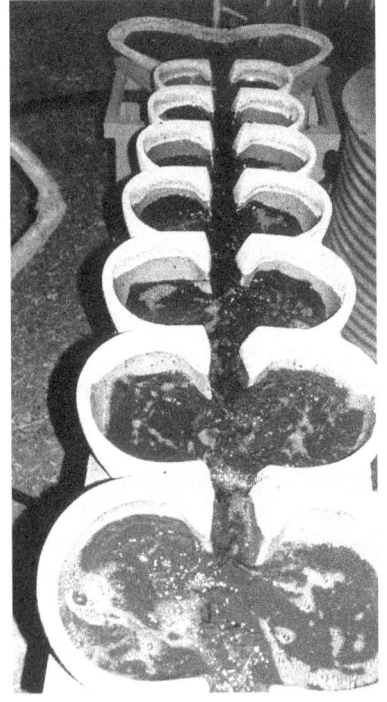

Their English inventor, the mathematician John Wilkes, observed natural patterns in nature and modelled his flowforms into the only water feature that moves water rhythmically in a figure-eight pattern on a horizontal plane, using precise geometric path curves incorporated into the design. The result is a 'rhythmical but dynamic visual and audio experience' that has soothing effects on stressed-out people and can be enlivening for the depressed, says a manufacturer.

Energising water with crystals

On a household level one can easily re-energise water to be more like 'living water'. Quartz crystals and other gemstones can help to positively re-program water. To do this, take a clear quartz crystal and cleanse it energetically under a little running water, or under the light of the full moon, or with your thoughts alone. Then 'program' it by holding it in your hand and asking it kindly to purify and energise the water it contacts, and thank it for doing so in advance. Place it in a jug of water and leave it there for 20 or 30 minutes minimum. Give it the taste test. When Sydney dowser Kathie Conover tried this she said that she was:

'impressed at the difference ... the water looks clear, 'bright' and tastes fine ... Another observation that fascinates me is the effect on cut flowers of being put in a jug of water containing a crystal. They seem to keep forever!'.

You can also mechanically energise water by shaking or stirring it around in a circle to create a vortex. Or stir it in a figure of eight and create both directions of vortex spin. You might like to try this by stirring bathwater with your hands or a crystal before getting into a bath.

Magnetising water

Restructuring of water can also be done with magnets. You might place a jug of water over a magnet and let it sit a while. Stir it around at the same time for an enhanced effect. Or stir a glass of water using a magnet, doing it swiftly enough for a vortex to form in the liquid. Keep this up for 30 seconds, then let the water stand for at least 6 minutes, then repeat the process a couple of times.

In Russia Dr Nikolai Yakovlev, of the Volga Research Institute of Hydraulic Engineering and Land Reclamation, reported in 1986 to have scientifically substantiated that magnetised water raises the microbiological activity of the soil and makes it easier for plants to take up nutrients. His team reported elevated yields of tomatoes and

cucumbers from doing this, at up to as much as 37 percent more crop. The method they used is to:
'obtain a strong horseshoe magnet, with a field strength above 1500 Gauss, and attach it to a hose that will propel ordinary water between the magnet's north and south poles at a rate of one metre a second, or thereabouts, before it reaches the plant.'

Using magnetised water in hot water jugs and appliances results in less 'scale' (mineral deposits) building up. A farmer told me that saline water becomes less harmful for crop irrigation when it is magnetised. Animals will also benefit from drinking it. Concrete is stronger and less cement is needed when magnetised water is used.

The power of prayer

The energy of our prayers has a similar effect to other 'psychic' or subtle-energy techniques. Scientific experiments using water as a medium for psychic energy has proven this, says EMR Labs, adding that prayer is:
[just as] '...effective as the energy from a person deliberately visualising a beneficial white light entering the water from his hands or eyes; and both have been proven via an electron microscope'.

In Japan the idea of remote healing by prayer was put to the test by Dr Masaru Emoto. Lake Biwa, a 670 square kilometre lake in Shiga prefecture, was marred by pollution with foul, murky water and toxic algal blooms each summer. The stench of putrifying algae could be smelt all around the area. But a group of people brought together by Dr Emoto had an amazing effect on it. The congregation of people went to the lake to use the power of *hado* (life force) and *kotodama* (the spirit of words).

Dr Emoto invited highly revered priest, 97-year-old Nobuo Shioya, to help preside over a mass prayer ceremony on the lake's shore, on 25th July 1999. The 350 participants chanted ten times the 'Great Declaration' as the rising sun came up:
'The eternal power of the universe has gathered itself to create a world with true and grand harmony.'

A month later the *Kyoto Shinbun* newspaper of 27th August 1999 had a very positive report:
'Every summer on Lake Biwa the foreign species of algae called

kokanada, that almost covers the lake as it flourishes abnormally, is practically out of sight this year. The complaints of foul odour which normally come in to Shiga prefecture have been nil this year, and the amount of algae cleaned up is a mere pittance'.

A dowser cleanses water

America's Raymon Grace is a modern pioneer in using dowsing and energetic methods of water purification and enhancement. It started when Grace was told about a contaminated spring in Arkansas. It had once been pure and its water bottled for sale, but it had since become polluted and undrinkable. Grace was asked to go there to see if anything could be done to cleanse the water. He said:

'*I work with what I refer to as the 'Spirit World' ... I started to communicate as best I could with some of the spirits of the land there and found that there were some Indian spirits still around that spring.*'

The spirits told him that they wanted him to purify the water, but didn't give him any suggestions on how to go about it. He decided to try communicating with the water itself and intuitively took his boots off and sat down with his feet in the water. The water gave him a message:

'*Everything has a frequency. Scramble the frequency of the pollutants of the water, and adjust them to the frequency of pure water.*'

To attempt to do this, Grace used his Dowsing System approach. He took out his pendulum and let it rotate over the water in a counter-clockwise spin. He explained:

'*Dowsing is just a way to focus intent, which is basically what a prayer is. All I ever do is ask. I just asked for a change in the frequency. I asked with respect ... [and] said a few words of gratitude to the Spirit of Water.*'

The water has since been tested by Nature Conservancy, who told the owner of the spring that it was one of the purest springs they had ever tested. Grace has concluded from his work that:

'*Water has a spirit which responds to respect ... the spirit of water would not go where there was greed. We have also found that the attitude of the landowner will affect the condition of the water on the land. In addition, curses on the land will also affect the water.*'

Nowadays Grace and the Water Project, based in Toronto, Canada, energise bottles of water and distribute them for free. Grace explains that:
'We have discovered that we can put a thoughtform into the water to cause it to energise all water that comes into contact with the energy field of the water. This was demonstrated by sitting a bottle of energised water on a table next to a bottle of plain water. The next morning the plain water had an energy field equal to the energised water. It has had some surprising side effects – my friend Jeannie energised a part of the Atlantic Ocean where she was swimming and 20 dolphins showed up and stayed all day!'

References

Water quality:
www.health.vic.gov.au/environment/downloads/your_private_drinking_water_supply.pdf
Caldicott, Helen, *Nuclear Power is not the Answer to Global Warming or Anything Else*, Melbourne University Press, Australia, 2006.
Australian Water Conservation and Reuse Research Program, Outcomes of Reviews and a New National Portfolio of Innovative Projects, Australian Water Association, 2004.
Keegan, Lyn & Keegan Gerald, *Healing Waters*, Berkley Books, USA, 1998.
'UN says number of ocean 'dead zones' rising fast' 19th October 2006, Source: Reuters, at www.alertnet.org/thenews/newsdesk/L19301317.htm
Weiss, Kenneth R., 'A Primeval Tide of Toxins', *The Los Angeles Times*, 30th July 2006.
Water Recycling in Australia, a review undertaken by the Australian Academy of Technological Sciences and Engineering, 2004.
Hartley-Hennessy, Therese, *Healing by Water or Drinking Sunlight and Oxygen – a new yet very ancient approach towards disease*, 1950, Essence of Health Publishing, South Africa.
Carseldine, Barbara, *Creating a Culture with a Reverence for Water*, self-published, Australia, 2003.
Faber, Scott, 'Sky rivers – atmospheric water-vapor transport', *Discover*, January 1994.
Topsfield, Jewel, 'Recying sewage should be last resort', *The Age*, 5th June 2007.

Rainwater Tanks a Viable Urban Water Solution, Australian Conservation Foundation, 16th April 2007.
Solar Water Purifier: website- www.solarwaterpurifier.com
Flowforms: www.virbelaflowforms.anth.org.uk.
Conover, Kathie, *NSW Dowsers Society Newsletter*, Feb. 1991.
'Magnetic touch in the garden', *Dowsers Club of South Australia newsletter*, April 1986.
Grace, Raymon, 'Water works', in *Dowsing Today*, UK, September 2004.
Salter, Prof. Stephen, 'Edinburgh Ducks', *New Scientist*, 7th November 2006.
Browne, Anthony, 'The spray turbine – How an egg-beater at sea could end drought and war', *The Times*, UK, 2nd December 2002.
CETO: at – www.carnegiecorp.com.au
Shiotani, *Jizairyoku 2*, Sunmark Press, Japan, 2000.

4.4 Water wisdom today

The Spirit of Water may be often ailing, but her powers can be boosted and restored. Healing the waters in a spirit of love and respect is something we can all do and it will help sustain our environment into the future. But where to start? In this last chapter we shall look at a range of traditions both old and new that can guide us to how we may better relate to water, and raise personal and planetary hado in the process.

Developing a spiritual relationship with water

The Finns and Saami of northern Europe show great sensitivity towards water with some beautiful traditions. They call the Spirit of Water *Mere-Ama*, *Vete-Ema* or *Mier-Iema*. A Sea Mother, she resides in fresh water too. The people honoured Her at ceremonies where they splashed each other with water. They sought the blessings of the Water Mother to assure good health and fertility for the people, livestock and crops. When a woman married and moved to a new home one of her first duties was to make herself known to the local Mere-Ama. She would go to the nearest stream and give offerings of bread and cheese, or cloth and thread, and wash her face or sprinkle herself with the waters.

I encourage people to revive this delightful tradition. Opening oneself to the spiritual dimensions of the natural world, I've discovered, allows all sorts of amazing things to happen. And I get a strong sense that the Spirit of Water really does want us to make these connections and that harmony can be restored or maintained with the healing and connecting powers of water and its compassionate consciousness.

A touching story illustrates this well. I met a woman a few years ago who lived alone, far from her original home. She had a good job and a nice home, but was lonely and loveless. A bush gully and stream were nearby and I introduced her to an important Water Mother spirit that I had discovered by dowsing there.

Some time afterwards she told me a wonderful story of what had happened after she had connected with that deva. By dowsing she was able to pinpoint the area where it was based and then enter into 'dialogue' with it. (Dowsing can provide answers to questions posed and help facilitate telepathic communications.)

A bit later a wonderful man came into her life. Unfortunately he had a debilitating, persistent skin problem that no cure could be found for. His hands were painfully cracked and bleeding. In desperation she asked for advice from the water deva, who told her to leave a bottle of water there under the moonlight for two nights. She did as instructed and he drank the water, one cupful each night before sleep. The psoriasis cleared, the cracks closed and he regained more use of his hands. Not long afterwards they were married and both their lives were transformed for the better!

People power at work

Land degradation in Australia has been rife, but many of the wounds are being healed. Across the country the enthusiastic, cohesive efforts of communities working together on a voluntary basis as Landcare groups has been saving and restoring whole landscapes. Their most successful strategies work from a catchment /watershed basis and as a result many waterways are being revived. Problems of weeds, erosion and declining bio-diversity are being tackled in a coordinated fashion, achieving vast amounts of tree planting and bush care.

While a lot of landcare work may involve using herbicide sprays to clean up weed-infested areas, or even bulldozing, it can be also be done with great sensitivity and even give people a psycho-spiritual boost in the process. I was trained in the Bradley method of bush regeneration. The Bradley sisters developed their thoughtful and systematic strategies around Sydney's exquisite bushland areas as they removed weeds gently by hand. The bush was being swallowed up by infestations of exotic plants that become monocultures, depriving wildlife of habitat. The Bradleys started to weed in the pristine bush areas and then worked towards the weedier bits. They went for a minimal disturbance approach and targeted upper catchment areas, to prevent weeds spreading rapidly down the waterway corridors.

In the mid-1980s I set to work on the weed problem in the Blue Mountains, New South Wales, with a small group of women who were volunteer bush regenerators. We also all had a common interest in meditation. We often combined both activities at the weed-infested sacred women's initiation ground of Minni Ha Ha Falls, in the north of Katoomba. The once 'bottomless' pool below the high waterfall was shallow with silt washed down from disturbances upstream. Prickly blackberries and dense thickets of Scotch broom choked the creek banks and most of the valley floor. It was a sad, neglected spot amidst the grandure of the sandstone escarpment country.

We concentrated on that special site, gradually removing a vast amount of weeds over time and allowing the natives to return, while enjoying the exercise, fresh air, sunshine and good company. Now when I visit I'm proud to see it as nature intended, with diverse vegetation and a scene of magnificent beauty!

At women's spirituality workshops that I ran 20 years ago I took groups of women there to frolic in the pool beneath the Falls and sing to the cascading waters, all naked of course!

Since those days the hado of Minni Ha Ha Falls has been profoundly uplifting. Aboriginal people still make pilgrimages to this highly sacred site.

Walking the Wimmera

In north-western Victoria the Wimmera River is the lifeblood and only river of the region, supplying water for wetlands of national and international significance as well as for stock, household use and crop irrigation. But the precious waterway has been gradually drying up over the last ten years. At Jeparit, which 25 years ago was a haven for fishermen, the river was dying from drought and accumulated levels of salt that had made it more salty than the sea, *The Weekly Times* (28th June 2006) reported. Parts of the lower Wimmera were below ground level and groundwater in the district had all gone saline too.

In March 2007 an important exercise in getting to know the river began. An initiative of the Elmhurst and Rainbow Landcare Groups, with support of the Wimmera Catchment Management Authority, it took this awareness to the whole community in the process. Participants of the two-week, 350-kilometre Mountains to Mallee Trek aimed to walk the length of the Wimmera 'as a tribute to the river and our waterways', as Rainbow Landcare secretary Heather Drendel put it.

It had been a big planning process to organise logistics and make connections with all the landowners along the way to gain their support and permission to walk on their land. On the first day of the trek 48 people joined in, but numbers fluctuated daily. Five members did the full trek, camping out each night. Others just joined in for short sections of the 'educational stroll', which included community functions en-route and a night walk across the dry Lake Hindmarsh.

The walk began in the hills near Elmshurt at the bush spring considered to be the river's source, on the north-west slopes of Mount Buangor. Here they filled a possum skin vessel, made by Wotjobaluk elders, with the spring water. At the end of the trek, at Lake Albacutya, they met again with locals, including the elders, and planted a eucalyptus tree, watering it from the spring water they had brought all the way from the river's source. A commemorative plaque was erected beside the tree.

Despite feeling sadness at the dry conditions, they were enthralled by the beauty of the river country and the wildlife they saw along the way. They got to hear the stories of sections of the river that they didn't know about and it was all documented on film. Wotjabaluk elders at Glenorchy, north-west of Stawell, welcomed them to their country and told the walkers the story of Purra, the giant kangaroo spirit, who began his Dreamtime journey creating the river and its chains of lakes in the area. Schoolchildren at Glenorchy commemorated the trek by depicting it in a special public mural that they had made.

A few weeks after the trek ended I asked Heather Drendel how she felt about her experience. She told me that it was a highlight of her life and she was obviously still on a high from it! She was buoyed by the way that people along the length of the river have become more positively focused on river care as a result, and she hopes that other rivers will also get to be walked this way. She said:

'I felt so priviliged to be one of the walkers. We learned so much. It was wonderful to see the landcare work that people have done along the way and to see the pride they take in the regeneration that's happening, in spite of the drought...

I still find it hard to sum up the experience, it was so huge.'

Healing the waters

Ilyhana Kennedy:
'The land cries out for our loving communion, as Indigenous peoples would once sing to all the waters! Our streams need mothering, nurturing and visiting with joy and respect. There is something special that I also do.
   I take sterilised clean bottles to the stream source in my locality. I ask Mother Earth for a little of the source water and take just a few bottles. I keep some for myself and use it to pour small quantities into bottled water that I drink each day. The rest is used to homeopathically dose the waterway downstream from the source, emptying a little at a time into different places, such as above and below a dam.
   (If you do this, always work with the same stream in the same catchment area and take care that you are not contravening any laws related to tampering with water storage and catchment areas.)'

Agnihotra ritual cleanses a sacred spring
by Parvati.

Messages from the spirit Orion are channelled by Parvati in Poland. In late 2005 Orion told Parvati to look into an ancient key sacred site at Karpacz that needed healing, and to perform Agnihotra fire there to give it a blessing. Parvati goes on:
   'When we were invited to give a talk on Agnihotra at the mountain village of Karpacz we knew we had several reasons for going. On the way, we met Ela who happened to also be a professional tour guide! As soon as we met, she informed us that there was an ancient site she wanted to take us to see. We looked at each other in amazement, as we had not met her before and here she was guiding us to the exact place we were sent to find!
   So we went to Saint Anne's shrine, which was a Catholic site that had been built on top of an ancient holy energy site. Beside it is a sacred spring which is considered to be miraculous water. According to legend it has healed many people over thousands of years. We were preparing to do a healing fire there with the idea of activating these healing energies. When we asked where it should be done, I listened to guidance that said, 'Trees hold the memories and the energies.' I related this to Ela

> *and she led us to a majestic 600 year old tree just a few metres away. We performed the fire under that tree and I received a message there.'*
>
> Orion told Parvati that the ancient site had once been a place of regular women's purification ceremonies involving fire, water and contact with elementals. He said that:
>
> *'What once was so alive is now being revived by many in various traditions, breaking barriers of silence and mystery. It is a time when the power of the ancients is being activated. Your work is done now.'*
>
> Parvati's message was confirmed that night:
> *'At the evening meeting we taught Agnihotra to an enthusiastic group of women who had gathered from Karpasz and the nearby city Jelena Gora. At the end of the talk, one woman announced that she was writing a book about 'magical Karpacz' and, without us ever telling anything about what we had just experienced earlier in the day, she began to speak, saying that long ago, before Christianity, on a group of rocks near the healing spring, women priestesses performed magical rituals for healing and prosperity on the full moon. These rituals were connected with the element of water. They created an energetic matrix and the natural spring was of healing water. However, it was not maintained. If only this place was activated (empowered), the energies would start again to be healing. If activated, the waters would once again give a powerful effect.'*

World day of love and thanks to water

25th July has been designated as a 'World Day of Love and Thanks to Water'. The suggestion is that people visit special water bodies to honour water each year on this day. Or, if participants can't go out, to honour a glass or bowl of water held in the hands. The group says:

'We have a vision that on this day, our Earth will be filled with beautiful golden/silver light of Love and Thanks that is flowing from the hearts of each and every one of us. Golden/silver light is the highest vibration in the range of visible light, and it will heal and cleanse all the water on Earth, be it water of the ocean or that of our own body.

Will you join us to say 'I love you' and 'I thank you' to all the water on Planet Earth and fill it with the highest vibration /hado of love and thanks that we can possibly experience?'

Tibetan Buddhist water rituals

Buddhism has rich traditions of maintaining spiritual harmony with all nature. Buddhism respects the animist traditions of places and its beautiful rituals for peace and harmony often feature water.

In December 2005 the Venerable Kirti Tsenshab Rinpoche went on a special mission to Antarctica to offer blessings to help stabilise the planet, to bless the world's water and to harmonise its weather. Rinpoche took holy water that had been blessed by the Dalai Lama, as well as water from all the oceans, from many lakes and rivers, plus sand from sacred sand mandalas. The mixing and unification of the many waters was done to bring about pacification of extremities in weather patterns.

In 2006 a Buddhist rain-making ritual was conducted west of Marong near Bendigo in Victoria. Monks from the Atisha Centre (where the Western world's largest stupa is being erected) gathered in bushland beside a dry watercourse for their rain ceremony. It had been one of the driest years on record, yet a few days later an isolated deluge of some 60 millimetres of rain fell, a friend who attended tells me.

Many ways to care for water

Water responds so well to vibrations of prayer, harmonious sounds and song, that expressions of kindness to water can be a joyful activity. Whether it is merely a sincere expression of gratitude to the rain, or the act of cleaning up rubbish at the local spring that you regularly visit, or an actual ritual at a desecrated spring with a circle of friends singing in harmony: it all helps to heal the rift between us, nature and Mother Water. We need to work on all levels in our multi-dimensional world. If we develop our own personal purity, clarity and vitality, we can be better catalysts for energetic improvements in our environment.

One might create a sacred water feature in the home or garden featuring special stones, plants or goldfish. It could be a solar–powered waterfall feature, designated as a home for water spirits, who would love to play in the moving waters! But whatever we do, what really counts are our intentions and actions.

Discover a local waterway, spring or pool that has been sacred to Indigenous people and approach it in a state of pilgrimage. Even if a special water place may never have been considered sacred (such as a modern artificial waterway in a park), you can choose to consider it sacred. It can be your connection point to the great Spirit of Water.

Be aware that Aboriginal women's sacred sites are often located near rivers and in wetlands. Sacred birthing sites, in particular, are found close to water. It is not appropriate for men to visit these places. Feelings of discomfort are often experienced when people intrude on sacred ground in violation of the strong cultural gender taboo there. It's best to go back at that point. But women may want to visit and enjoy the sacred women's places. Take a picnic!

Go to your nearest spring/pool/beach/dam/waterway and perhaps take a flower or leaf or feather that you might find on the walk there, to offer as a gift to the Spirit of Water. Place it on a special rock or high energy spot and with it convey your heartfelt love and gratitude.

In the process we cultivate these qualities in ourselves. The verbalisation of deep love and thanks to water can have powerful resonances throughout nature, as well as to the waters of our own being. So sing and talk to the waters in and around you, even if it is only the water in your glass or shower, or the water with which you wash your hands or water the garden.

In Japan in the early 1990s Masami Saionji introduced the idea of spending a little time each day to voice affirmations of gratitude to all nature. He wrote the followings words for people to express their gratitude to water:

'On behalf of humanity, we thank you, divinities who govern water.
We thank you, dear water. Without you, we could not live.
Please forgive the foolishness of defiling you with the boundless
greed of human beings.
We heartily thank you for your precious existence and functions.
May peace prevail on Earth.
May the divine missions of water be accomplished.'

Photo: Pioneer Women's Fountain, King Park, Perth, Western Australia. A sacred site both old and new and a perfect place for quiet reflection and meditation.

References

Monaghan, Patricia, *The Book of Goddesses and Heroines*, Llewellyn, USA, 1981.
Wimmera Catchment Management Authority –
www.wcma.vic.gov.au/index.php/waterways
Rinpoche, Kyabje Lama Zopa, *Who is Kirti Tsenshab Rinpoche?*, 2006, on-line.
The Project of Love and Thanks to Water website
Parvati, *Orion Transmissions*, December 2005, via Homa Therapy Association of Australia, P.O. Box 68 Cessnock NSW 2325, email – omshreedham@optusnet.com.au
Saionji, Masami, *Gratitude to Nature*, Byakko Shinko Kai, Japan, 2001.

Python Press Books

Backyard Poultry - Naturally

From housing to feeding, from selection to breeding, from pets to production and from the best lookers to the best layers, this book covers everything the backyard farmer needs to know about poultry husbandry - including preventative and curative herbal medicines and homeopathics, plus permaculture design for productive poultry pens. Useful plant profiles, feed recipes, simple natural remedies and more.

This is Australia's best selling poultry book!
The Reviews:
"The poultry health section is the best I've seen." Eve Sinton, Permaculture International Journal.

"A wonderful resource! Alanna Moore has provided poultry enthusiasts with all the information they need to raise healthy poultry without using harmful chemicals." Megg Miller, Grass Roots magazine.

"An interesting and worthwhile book that will no doubt have a lot of appeal for the amateur or part-time farmer." Kerry Lonergan, Landline, ABC TV

Stone Age Farming
- eco-agriculture for the 21st century

An exploration of energetic growing techniques that are in harmony with nature. Ancient and new ideas about the energies of rocks are explored for practical application in the farm and garden. The use of dowsing for soil testing and other analysis, as well as the advanced dowsing system known as radionics are described.

Looking to ancient Irish mysteries, such as the 1000 year old Irish Round Towers, practical uses for esoteric knowledge include modern Towers of Power that are being constructed for enhanced plant growth and animal wellbeing. These paramagnetic antennas work by harvesting and re-radiating intensified atmospheric magnetism around them.

What reviewers have said:
"Simply fabulous!" Maurice Finkel, Health and Healing
"Quite fantastic." Roberta Britt, Canadian Quester Journal
"Clear, lucid and practical" Tom Graves
"A classic" Radionics Network

Python Press Books

Sensitive Permaculture
- Cultivating the way of the sacred Earth

This 2009 book explores the living energies of the land and how to sensitively connect with them. Positive and joyful, it draws on the indigenous wisdom of Australasia, Ireland and elsewhere, combining the insights of geomancy and geobiology with eco-smart permaculture design to offer an exciting new paradigm for sustainable living. It includes the authors experiences of negotiating with fairy beings and dragon spirits over land use in Australia and Ireland and, as in the co-creative gardening at Scotland's Findhorn community, there's a giant cabbage in there too!

Praise for Sensitive Permaculture:
"A delight to read" Callie
"...Hard to put down" Celia,
Permaculture Association of Tasmania

"You make permaculture so easy and alive---and sweet" Joy, Taiwan

"A very practical and thoughtful guide for the eco-spiritual gardener, bringing awareness to the invisible dimensions of our landscape"
Rainbow News, New Zealand

"An adventure in magical and practical Earth awareness" Nexus magazine

The Magic of Menhirs & Circles of Stone

Discover the world of standing stones (also known as menhirs and megaliths), stone circles, medicine wheels and labyrinths. This small book is rich in practical insights for creating ritual stone arrangements in your own backyard. Where to construct them and how to utilise them for improved Earth harmony and personal benefits. Includes recent energetic investigations and discoveries by dowsers and clairvoyants around the world.

"Have you ever wanted a sacred site in your own backyard? This book will help you to achieve that dream"

Don McLeod, 'Silver Wheel', Sth Australia.

Python Press Books

Divining Earth Spirit

A comprehensive guide to geomancy and geobiology, updated in 2010.

"*Highly recommended*"
Glastonbell Vol. 5 No 4

"This book is a classic for anyone wanting to get involved with Earth healing. It contains information by the bucketload... The research that has gone into this book is incredible and no doubt will stir you into wanting to use it yourself" Radionics Network Vol. 2 No.6

"Excellent reference book"
Don McLeod, Silver Wheel

"Love of the topic clearly shows, as Moore brings clarity and a sense of the necessity of personal involvement and engagement with the Earth. The great advantage of Moore's book is in its detailing all the salient aspects of Earth Spirit phenomena....all covered succinctly and with precision... the perfect introduction to the topic,"
Esoterica magazine, No. 4, 1995

Water Spirits of the World
- from nymphs to nixies, serpents to sirens

A follow-on from The Wisdom of Water, this book delves into even deeper esoteric aspects of water and its spiritual denizens. For example, the universal significance of the serpent in global traditions reveals this spirit of water as fundamental to the origins of spirituality, a clue to its denigration in later religions of historical times. Water spirits are still to be found, even in urban landscapes, and Alanna's experiences with them show that the ancient myths are based on the spiritual reality of nature.

What readers have said of this book:
"A wonderful resource book"
Morgana, 'Wiccan Rede' 2009, Holland

"This joyful travelogue of water spirits around the world has been a journey inspired by love"
Anne Guest, 'Gatekeeper' no. 26, UK

"A comprehensive collection of information and a rich insight into the world of water spirits ... including some wonderful stories of encounters with water spirits ... well researched and informative"
Martha Heeren, Dowsers Society of NSW newsletter, April 2009.

Geomantica Films

Geomantica presents information about dowsing, geomancy and esoteric approaches to sustainable living and Earth harmony.
Nineteen films by Alanna Moore are currently available, including:

Grassroots Solutions to Soil Salinity
38 minutes

Despite government scientists saying it can't work, some farmers in Australia are successfully tackling soil salinity and land degradation. This film shows two farms that have been saved by installing interceptor banks to divert sub-surface water flows. See dowsing as applied in the design of these banks and meet water scientist Rob Gourlay who shows how the official theory of a 'rising water table' as the cause of soil salinity is a fallacy that prevents remediation of the problem.

The Sacred World of Water
38 minutes

This film explores the many curious characteristics of water and it's traditionally sacred dimensions. From folklore to spiritual water divining, with water rituals and some spectacular sacred water sites in Australasia, England and Ireland.

Praise for this film:
'This film takes you on a fact-filled journey ... [with] a variety of purification rituals, the persisting legends around wells, springs and aspects of The Great Mother. ... The overall theme of this film is the need to respect and care for this Element .... Alanna Moore's films are unique, individual and idiosyncratic. They are almost the documentary equivalent of the Japanese poetry form of haiku; short and focussed, with every component being Very Important. You have to concentrate, almost meditate, as you watch. I can only repeat that if you like visual imagery and verbal guides, then these films are an enjoyable way to take a break from the written word. If you seek stimulation, inspiration, or even a starting point for your personal spiritual quest, then you may well find that nugget in one of Alanna Moore's films.'
Jilli Roberts, Pagan Times, Spring 2006

Geomantica

PO Box 929 Castlemaine
3450 Vic Australia
email: info@geomantica.com
Free geomancy magazine & information,
mail order sales of Geomantica Films
and more at:
www.geomantica.com

Python Press Books

Distributed in Australasia,
Europe and North America
including Amazon and other on-line suppliers
Also from Geomantica by mail order
(but check availability first)
For further information:
www.pythonpress.com
email: pythonpress@gmail.com